Analytical Methods in Oceanography

Analytical Methods in Oceanography

Thomas R. P. Gibb, Jr., EDITOR

Tufts University

A symposium sponsored
by the Division of
Analytical Chemistry
at the 168th Meeting
of the American Chemical
Society, Atlantic City,
N.J., September
10–11, 1974

ADVANCES IN CHEMISTRY SERIES 147

AMERICAN CHEMICAL SOCIETY
WASHINGTON, D. C.   1975

Library of Congress CIP Data

Analytical methods in oceanography.
    (Advances in chemistry series; 147 ISSN 0065-2393)

    Includes bibliographical references and index.

    1. Chemical oceanography—Technique—Congresses.
    I. Gibb, Thomas Robinson Pirie. II. American Chemi-
cal Society. Division of Analytical Chemistry. III. Series:
Advances in chemistry series; 147.

QD1.A355 no. 147 [GC110] 540'.8s [551.4'601'028]
                                                          75-41463
ISBN 0-8412-0245-1   ADCSAJ 147 1-238

Advances in Chemistry Series

Robert F. Gould, *Editor*

FOREWORD

ADVANCES IN CHEMISTRY SERIES was founded in 1949 by the American Chemical Society as an outlet for symposia and collections of data in special areas of topical interest that could not be accommodated in the Society's journals. It provides a medium for symposia that would otherwise be fragmented, their papers distributed among several journals or not published at all. Papers are refereed critically according to ACS editorial standards and receive the careful attention and processing characteristic of ACS publications. Papers published in ADVANCES IN CHEMISTRY SERIES are original contributions not published elsewhere in whole or major part and include reports of research as well as reviews since symposia may embrace both types of presentation.

CONTENTS

PREFACE

This book is based on a symposium "Chemical Methods in Oceanography" which arose from an obvious suggestion I made a year or so ago that chemical oceanographers and analytical chemists should talk to each other more. The lamentable fact is that analytical chemists get seasick and that oceanographers have to cope with shipboard conditions which include unthinkable per diem costs, varying acceleration in all possible directions, and an unexceptional lack of items not brought aboard in the first place.

These facts of marine life cause oceanographers to grow beards and become taciturn. Their work is specialized, and they tend to publish in journals which are either benthic or, at best, pelagic—and largely unread by land-based analytical chemists.

The converse is not true, marine chemists are well informed of the latest tools of analytical chemistry and often use these new methods at a high level of sophistication. This volume is not designed to inform sea-going chemists of recent advances by shore-based chemists, although one or two of the chapters may indeed do this. Rather its purpose is to acquaint land-locked chemists with the accomplishments and problems of marine chemists so that the accomplishments will be more widely honored and the problem solving shared. There are enough problems for all of us as Dr. Hume points out in his keynote address.

There is another purpose—to let marine chemists take a broader look at what they are doing. Analytical chemistry is more than a tool; it is a large body of knowledge and a discipline. A good physical chemist, for example, can be a poor analytical chemist. The science of finding out what is present and how much has a touch of artistry in it.

The marine chemist is faced by extremely difficult problems, and it is probably correct to state that some of these are well beyond the state of the art at the present time. If an oceanographic theory is to be based on 6000 analyses for sodium glutamate, it would be well to know that those analyses are significant. They could be high every time the crew had chow mein for dinner. Or they could be low because there were no fingerprints on the sampler. Or the shipboard routine of sampling, storing, and analyzing just wasn't good enough.

If we were to believe data reported for trace metals in seawater and fish over the years since 1900, it follows that fish must be consuming trace metals and removing them rapidly from the oceans! A plot of the heavy metals content of seawater *vs.* years shows a steep asymptotic decline whereas the trace metals content of fish shows a startling but irregular increase. These results are, however, susceptible to another less fishy but equally embarrassing interpretation: until recently all seawater samples were contaminated by the sampler, and all our analyses were high, sometimes by orders of magnitude! This skeleton in our closet is about to be buried at sea, and perhaps someone will read its burial service from the pages of this book.

We have made other kinds of errors in our analytical determinations. For instance many of us have been measuring the pH of seawater using equipment calibrated by $0.050M$ potassium hydrogen phthalate. Have we really been measuring pH? The answer is no, and the meaning of our measurements is just beginning to be understood and with it the effects of salinity, temperature, and pressure.

It is my hope that this book will remind all of us that we have much to learn about the analytical chemistry of the oceans. I think it may help to delineate some of the major areas where work is needed and that it introduces some rather specialized, even exotic, areas that are quite fascinating.

I should like to thank the authors, the Analytical Division of the American Chemical Society, and the editor and staff of the Advances in Chemistry Series. They did all the work and should get all the credit.

THOMAS R. P. GIBB, JR.

Tufts University
Medford, Mass.
February 1975

Fundamental Problems in Oceanographic Analysis

DAVID N. HUME

Department of Chemistry, Massachusetts Institute of Technology, Cambridge, Mass. 02139

The success of studies on ocean chemistry usually hinges on the availability of adequate analytical methodology and expertise. The nature of the chemical problems involved; the size and heterogeneity of the ocean; the extremes of concentration, composition, temperature, and pressure; and the dynamic interactions at its interfaces with the lithosphere, atmosphere, and biosphere create challenging and often unique analytical problems. Current analytical methods are limited, and improved capability in referee, routine, monitoring, and in situ analyses is needed. Valid sampling, the use of reliable standards in calibration, and the collaborative evaluation of methods under real-world conditions are important.

The chemistry of the oceans involves a vast number of intricately and intimately interrelated systems and reactions. To be understood, many of these require precise and accurate analytical data, which is often unattainable by conventional means. This symposium has brought together oceanographers and analytical chemists for meaningful communication. We hope that the oceanographers can give an overview of the fundamental chemical problems of the ocean in terms of the equilibrium, dynamic, biological, hydrological, geological, and meteorological systems involved and that the chemists can speak cogently in terms of what approaches may be effective in generating valid data under the unusual and extreme conditions which characterize the natural ocean. In short, we will discuss what ocean chemistry is all about, what questions need answering, what the difficulties are, and what analytical chemists may be able to do to help.

When one considers the size of the ocean—its 10^{21} l. of water covers 71% of the earth's surface—and its importance as the climatic mediator, as the source of food and materials, and as the ultimate sink for many of our pollutants, it is surprising how little is actually known about it. Only in very recent years has the volume of ocean research, chemical and otherwise, started increasing rapidly. With this increase has come an awareness of the inadequacy of our methodology.

The Ocean as an Object and a System

The ocean is a large mass of water containing about 3.5% salt in a fairly uniform mixture. It is layered with respect to temperature, being warm enough at the surface in many places for swimming but close to 4°C throughout most of its depths. There are various horizontal and vertical currents, which produce gradual mixing. From the standpoint of a chemist it is almost a static system because the mixing time is about 1600 years, although this constitutes rapid turnover to a geologist who consequently views it as a well mixed dynamic system. While there are local variations in total salinity because of evaporation, rainfall, and input from rivers, the relative proportions of the major constituents are remarkably uniform (Table I) (1). The bulk of the dissolved salts (> 99.99%) is made up of only 11 elements which vary only slightly in their relative amounts. Most of the rest of the dissolved salts consist of the minor constituents (Table II) (2) which vary considerably in their relative and absolute amounts, largely as the result of biological activity. What is left is a little bit of everything else in the periodic table at concentrations of a few parts per billion and less.

It would be a considerable challenge just to establish accurately the composition and chemical properties of an isolated sample of seawater, but the sea is not isolated. It has three important interfaces with its environment: the atmosphere, the lithosphere, and the biosphere. Each of these interfaces involves active transport of matter and energy, each

Table I.   Concentrations of the Major Nonvolatile Constituents of 35‰ Salinity Seawater

Constituent	Concentration (mg/kg)	Constituent	Concentration (mg/kg)
Chloride	19,350	Bicarbonate	142
Sodium	10,760	Bromide	67
Sulfate	2,710	Strontium	8
Magnesium	1,290	Boron	5
Calcium	413	Fluoride	1
Potassium	387		

Table II. Approximate Concentrations of the Minor Nonvolatile Constituents of 35‰ Salinity Seawater[a]

Constituent	Concentration ($\mu g/kg$)	Constituent	Concentration ($\mu g/kg$)
Silicon	~3000	Barium	30
Nitrogen	~1000	Aluminum	~10
Lithium	170	Iron	~10
Rubidium	120	Zinc	~10
Potassium	~70	Molybdenum	~10
Iodine	60		

[a] All remaining elements <10 ppb

has special properties of its own, and each affects the others. Let us look at some of the problems which result.

The Water–Sediment Interface

First, there is the matter of marine geochemistry. Quite apart from the problem of what is actually present in terms of water and sediment composition (which is by no means an exhausted topic) there are two related questions: How did the sea get its present composition? Is the sea in an equilibrium condition, in a steady state, or in the process of change? As a rough approximation the origin of seawater can be described in terms of the reaction of 600 g of primary rock reacting with 1 kg of volatiles (sulfur dioxide, hydrochloric acid, carbon dioxide, and water) to yield 1 l. of seawater, 3 l. of air, and 600 g of sediments. With the proper assumptions, this gives a plausible material balance, the details of which can only be established with a good deal more data. The leaching and dissolution of rocks and soil continue. Every year rain and juvenile water goes down the rivers of the world carrying a great burden of dissolved and suspended matter. The composition of the rivers differs drastically from that of the sea, yet the sea does not seem to have changed its composition in some millions of years. What are the processes which keep things in balance? If reactions are taking place, for example, at the sediment interface, very small changes in the relatively large concentrations of the major components will result. To detect these differences and measure them with any certainty is important geologically and difficult analytically. Of particular interest is the composition of the interstitial pore water in the sediments below the bottom of the ocean. This is water which exchanges and mixes very slowly with ordinary bottom water, and its composition may be the result of centuries of equilibration. To sample such water reliably and without contamination is a particularly challenging task. It can mean sending a sampler down several miles to get a few milliliters by remote control under a pressure of 1000 atm and then getting it back up again unaltered in the process. Recent work using improved sampling and analysis techniques indicates that some elements previously thought to be concentrated in pore water are actually depleted, or vice versa, which suggests the need for a 180 degree adjustment of some geochemical theories of sedimentary processes (3).

Even apparently straightforward reactions develop unexpected characteristics when allowed the freedom of the seas. It has long been known, for example, that some planktonic organisms withdraw calcium and carbon dioxide from the water to grow hard-parts of calcium carbonate. These skeletal remains sink toward the bottom after the organisms die, and in fact large areas of the ocean bottom are covered with thick layers of sediments rich in them. It would seem obvious that surface waters, where the growth takes place, should be depleted in calcium, and deep water in contact with the excess solid calcium carbonate should be saturated. In actual fact the reverse is true. Deep water is undersaturated, and surface water is supersaturated, sometimes by more than 200%. What is involved is a fascinating array of complicating factors: the effects of temperature and pressure on equilibrium constants, the influence of many things such as adsorbed films on the kinetics of the reactions taking place, and the changing depth of sediments through the action of plate tectonics. The final picture is still far from clear. Even as simple a thing as why the equilibrium pH of seawater is characteristically around 8.3 is not settled. One theory is that the sea is a big pool of bicarbonate buffer. Another is that the governing process is an ion exchange process with the sediments in which potassium and hydrogen are the active members and that the carbonate–bicarbonate–carbon dioxide system is more of an indicator than a buffer (4). Actually relatively little is known, compared with what is needed, about the kinetics and equilibria of sediment–water reactions under ocean bottom conditions.

The Water–Atmosphere Interface

Returning to the surface, all sorts of activities take place at the air–sea interface. There is active transport of gases—O_2, CO_2, N_2, H_2O vapor, CO, N_2O, CH_4, and He, for example—in both directions. Carbon monoxide is an interesting example. Until very recently it was assumed that the sea was the principal sink for the carbon monoxide produced by man's civilized activities. However, careful analysis of oceanic and continental surface air disclosed that carbon monoxide was coming out of, not going into the sea, and that it represented the largest source of that gas. It is presumed to be of biological origin, but the actual source is not known (5).

The transport of water across the sea surface is particularly important to us because that is the source of most of our rain. Along with the water are carried rather appreciable amounts of salt. Unfortunately, the simple explanation of spray from waves neither accounts for the quantity nor the composition of the salt, which turns out to be appreciably different from that in bulk seawater. The explanation seems to lie in transport by bursting of microbubbles, the solute composition of which can be predicted to deviate in the observed direction by invoking the proper sophisticated theory of liquid water structure. The composition of the microscopically thin surface layer of the ocean, in which surface-active substances of natural and artificial origin collect and form films, actively affects transport of water and gases. Also of importance here is the effect of pollutants such as petroleum products and lipidophyllic substances from sewage, not only on transport processes but on biological activity which is concentrated near the ocean surface.

The Water–Biosphere Interface

This brings us to the bio–water interface and the subject of life in the sea. The sea, like the land, has a bioeconomy based on producing plants and consuming animals. Unlike the land the primary food production in the sea is nearly all done by microscopic plant life—the phytoplankton. Microscopic plants are eaten by microscopic animals, the zooplankton, which are eaten by increasingly larger animals up to sharks, men, and other predators. The factors governing the productivity of seawater are much the same as those governing the productivity of land—availability of the nutrient elements (mainly nitrogen and phosphorus, to a lesser extent silicon) and essential trace elements (Cu, Mo, Fe, Mn, Co, etc.). Many of these are present in seawater at very low levels and are readily depleted. Large areas of the ocean are, in fact, the aquatic equivalent of barren deserts because the nutrients in the surface zone are exhausted, and really high productivity is maintained only where nutrient-rich bottom water wells up to replace the supply. The study of the interrelationships between productivity and the concentration and speciation of essential trace elements calls for analytical methods of higher sensitivity and specificity than are yet dependably available.

Very little detail is known about the nitrogen and phosphorus cycles. Dissolved orthophosphate is taken up in plant growth and returned eventually after serving in cell fluids and tissues. It appears to be released primarily in the form of organophosphorus compounds, and at certain seasons of the year more than half the phosphorus in the water is described as "dissolved organic phosphorus." Yet as of a year or so ago when I last checked the literature on the subject, not one organic phosphorus compound in seawater had been conclusively identified. Here is an interesting analytical challenge. Indeed, all oceanic organic chemistry is a challenge. Seawater in general contains around 3 mg/l. of "total organic carbon," which seems to consist of a bit of every stable organic compound that exists together with quite a few of the more transitory ones. Considerable amounts appear to be in the form of difficult-to-degrade natural polymers such as humic substances and chitin. Hydrocarbons are found ranging from macromolecules to methane. Clearly a capacity to establish the detailed organic composition of a seawater sample would be a tremendous help in working out what was going on in the ocean, but present methods fall far short of that goal. It is a sad fact, as will be shown in one of the papers in this symposium, that even the standard method for determining total organic carbon is open to serious question.

Limitations of Current Analytical Methodology

It must be admitted that every analytical method is open to question at some level. Unfortunately, for many determinations in seawater, that level is right in the range of normal composition and conditions. Trace element analysis is a good example. A great deal of work has been done to measure heavy metals in seawater at their natural parts per billion levels. The results indicate considerable variation with location and depth. This is of great biological and geological significance, if true. Yet interlaboratory comparisons of results on the same samples frequently show shocking disagreement. Lead is a particularly well documented case in point. In a recent intercalibration study, several laboratories analyzed the same carefully taken and preserved seawater sample. Each worker was experienced and considered expert in his technique. He was aware of the discrepancies which had characterized previous intercomparisons and had been alerted to the known sources of error. Nevertheless, the reported averages obtained by anodic stripping voltammetry or atomic absorption ranged from 50 to 1300 ng/kg. The accepted value, obtained by isotope dilution mass spectrometry with heroic efforts to avoid contamination, was 14 ± 3 ng/kg (6). While this is an extreme example, it is by no means unique. Even the most sanguine optimist, on reviewing the results of trace level intercalibration studies, would have to concede that a large fraction of the trace element concentrations reported in the literature must be essentially worthless. Close examination of the performance of the standard methods for determining the major seawater constituents suggests further that many of the conclusions drawn from small apparent differences in those concentrations may also be illusory.

Present Needs and Future Directions

It is evident that there is an urgent need for new and improved analytical methodology for ocean research. These methods must have genuine overall precision and accuracy high enough to provide the discrimination necessary for answering the questions under study. Several kinds of methods are needed such as referee methods which are used when the right answer is essential regardless of time or cost. They validate the practical methods which will be applied routinely: methods simple and sturdy enough to be used on a ship and rapid and inexpensive enough to process the large number of samples produced in a survey. Also needed are monitoring methods which can be automated for unattended operation. Finally, methods by which analyses can be done *in situ*—even under the extreme conditions of temperature, pressure, and remoteness which are characteristic of the deep ocean—are called for if certain types of problems are to be studied.

However, determination methods are not sufficient in themselves. Traditionally academic analytical chemists have concentrated on developing measurement methods while tending to ignore the equally important but less glamorous problems of sampling. Even to define what a valid sample should be is no trivial matter. Ought it to include particulate matter or not? If not, how can it be separated, and, for that matter, what criterion distinguishes small particles from large non-particles? Rational, reliable, and practical methods for sampling bulk water, interstitial pore water, the surface layer, and the sediments are needed. Once obtained, these samples must be stored, sub-sampled, and processed using convenient and practical nondestructive and noninterfering protocols which, for the most part, do not exist yet. The effort put into analytical methodology and measurement is wasted if the sample loses its validity or integrity. That the analytical effort is not all that is wasted is evident when one considers that taking a sample from the ocean depths may involve holding a ship, with an operating cost of four figures per day, stationary for 4–8 hr. Improvement of sampling techniques and elimination of sampling by development of *in situ* analyzers are projects of the highest priority.

Another unglamorous, but vitally important, aspect of oceanographic analysis is the area of calibration and standards. Standard reference materials appropriate for most of the constituents in seawater simply do not exist, and considering the low concentration levels usually involved and the unstable nature of natural seawater samples, the way to approach the problem is by no means obvious. Calibration of laboratory methods and instruments is difficult enough but is nothing like the challenge of calibrating sensors designed for measurement *in situ* 6 mi. down and under 1000 atm pressure.

Clearly there are problems enough to keep many analytical chemists occupied for a long time. I want to emphasize, however, that working on these problems in the abstract is not likely to be productive. The methodology which is produced must work reliably and conveniently in the hands of the ultimate users and be applicable to real problems, or it really isn't much good. Rigorous evaluation and intercomparison of results obtained by different users on the same samples is an essential step in the development of any method which is to be accepted as standard. This means communication and cooperation between investigators, particularly between those of us who consider ourselves analytical chemists and those of us who consider ourselves chemical oceanographers.

Literature Cited

1. Culkin, F., "Chemical Oceanography," J. P. Riley and G. Skirrow, Eds., Vol. 1, p. 121, Academic, 1965.
2. Goldberg, E. D., "Chemical Oceanography," J. P. Riley and G. Skirrow, Eds., Vol. 1, p. 163, Academic, 1965.
3. Sayles, F. L., *et al., Science* (1973) **181**, 154.
4. Sillén, L. G., *Svensk Kemisk Tids.* (1963) **75**, 161.
5. Swinnerton, J. W., Linnenbom, V. J., Lamontagne, R. A., *Science* (1970) **167**, 984.
6. *Marine Chem.* (1974) **2**, 69.

RECEIVED January 3, 1975

Trace Metal Contamination by Oceanographic Samplers

A Comparison of Various Niskin Samplers and a Pumping System

DOUGLAS A. SEGAR and GEORGE A. BERBERIAN

National Oceanic and Atmospheric Administration, Atlantic Oceanographic and Meteorological Laboratories, 15 Rickenbacker Causeway, Miami, Fla. 33149

Simultaneous collections were made of water and suspended sediments by an Inter-Ocean pump sampling system, Niskin bottles with internal rubber closures, Niskin bottles with Teflon-coated coil springs, and newly designed Niskin bottles without internal closures. The samples were examined for contamination, particularly of trace metals. All the sampling systems led to some metal contamination of the samples except the new Niskin bottles. However, each system was found acceptable for nutrient analysis. Because of its novel closure, the newly designed Niskin bottle is convenient for suspended material analysis. Samples may be filtered directly from the sampling bottle without contacting the atmosphere.

The analysis of seawater for trace components, particularly metals, is a notoriously difficult task. Sophisticated and expensive analytical instruments are used, and many difficulties are incurred in the procedure. Considerable efforts have been expended on analytical problems (*1, 2*) and problems of sample contamination from storage bottles and reagents used for analysis (*3*). However, little attention has been paid to the problem of sample contamination during the process of obtaining the seawater sample from the ocean. Solving the analysis problems is of little use if uncontaminated samples cannot be obtained. In a separate paper in this volume, a new analytical technique is described, which eliminates the worst of the analytical problems for several elements (4). These new methods were developed for use in the National Oceanic and Atmospheric Administration's (NOAA) Marine Ecosystem Analysis (MESA) Project and for studies of water column chemistry over mid-ocean ridges. This paper describes a parallel effort to test sampling systems for trace metal contamination and to develop a sampling procedure which minimizes contamination.

Sampling Systems

Two basically different systems, pumps and sampling bottles, are used to obtain seawater samples. Bottles such as the Nansen, Niskin, National Institute of Oceanography (NIO), and Van Dorn samplers have been used extensively throughout the world ocean. Pumping systems have been used less frequently, usually only in shallow water and when large volume samples or continuous profiles are required. A review of the literature reveals that little has been published regarding contamination of samples from sampling bottles. Cooper (5), in reviewing the problems of seawater sampling, pointed out that contamination, particularly from metal and rubber components of the samplers, was a serious problem. Subsequently, sampling bottles made from polyvinyl chloride (PVC) were almost universally adopted for trace metal analysis. Unfortunately, the standard versions of such sampling bottles all use either rubber end caps or rubber "springs" inside the bottle as a closure mechanism. The most widely used sampler, the Niskin bottle, has silicone internal rubber closures that pass through the sampler and remain in contact with the sample after the bottle is closed. Such silicone rubber is reported to give off considerable quantities of zinc to solution. In addition, the rubber, particularly when new, gives off large quantities of particles to the water sample during the closing process (6). To solve this problem many researchers adopted a Teflon-coated stainless steel coil spring to replace the rubber. Unfortunately, no investigation of the contamination generated from this spring has been reported. However, Teflon is known to be porous, particularly in thin coatings or films, and the coatings themselves tend to crack with even very limited use.

Faced with these sampler problems, a new design of PVC Niskin bottle was constructed which eliminated the internal closures (7). This bottle (Figures 1 and 2) is designed to be closed from the outside. The top and bottom plugs are both loaded at the top of the bottle before lowering. Upon triggering, the bottom plug free falls within the bottle until it closes the bottom end of the bottle by resting on an O-ring seal. The top plug is pulled into place by three tensioned latex rubber bands outside the bottle and then locked in place by three spring-loaded cams.

Figure 1. Top drop Niskin bottle closed   *Figure 2. Top drop Niskin bottle loaded*

The sample thus comes in contact only with PVC during its time in the sampler.

The pump sampling system used in this study was an Inter-Ocean OSEAS system (7) which features a submersible, multi-stage axial flow pump. All metal parts of this system are Teflon coated so that samples may be collected for trace metal analysis.

Comparisons of Sampling Systems

Upon acquisition of the Inter-Ocean pump sampling system, a series of seawater samples was collected from New York Bight at 10 m depth simultaneously with the pump and with Niskin 10-l. samplers with internal rubber closures. The samples were filtered through 0.4μ Nucleopore filters. One aliquot was frozen for subsequent analysis of nitrate, nitrite, phosphate, and silicate by standard automated techniques. A second aliquot was acidified to pH 1 with silica-distilled, concentrated nitric acid and was subsequently analyzed by flameless atom reservoir atomic absorption spectrophotometry, both by direct injection (4) and after extraction of the metal pyrollidine dithiocarbamates with methyl isobutyl

Table I.   Comparison of Pump and

	Phosphate (µg at/l)		Nitrate (µg at/l)		Nitrite (µg at/l)		Silicate (µg at/l)	
Station	Pump	Niskin	Pump	Niskin	Pump	Niskin	Pump	Niskin
S								
6	0.70	0.69	3.0	2.9	0.21	0.26	<0.2	<0.2
8	1.2	1.4	7.8	9.6	0.94	1.1	5.2	6.3
12	0.64	0.58	2.6	2.2	0.14	0.10	0.2	<0.2
13	0.92	0.76	3.1	3.0	0.30	0.30	<0.2	<0.2
22	0.78	0.74	3.5	3.5	0.50	0.51	0.2	0.2
25	0.63	0.64	2.6	2.5	0.28	0.28	0.5	0.4

ᵃ Niskin bottle with internal closures

ketone (9). The results of these analyses are shown in Table I. It is readily apparent either that no nutrient contamination or loss occurs with either sampling system or that any contamination or loss is identical in each system. This second possibility is unlikely because of the considerable differences in the two systems. The same conclusion—that neither sampler contaminates the sample—may be reached for manganese. However, for copper, to a lesser extent for nickel, and possibly for iron, the pump system contaminated the sample.

In light of the above data, use of the pump system was discontinued, and a Niskin Rosette sampler (Figure 3) was used to perform a com-

Figure 3.   Rosette sampler with 10-l. top drop bottles

Niskin Bottle[a] and Collected Samples[b]

Nickel (μg/kg)		Manganese (μg/kg)		Iron (μg/kg)		Copper (μg/kg)	
Pump	Niskin	Pump	Niskin	Pump	Niskin	Pump	Niskin
4.7	4.6	6.1	5.9	110	115	17.4	3.8
2.8	2.0	3.5	1.6	412	131	17.6	2.3
2.4	1.9	0.57	0.59	43	32	15.3	1.4
1.4	0.7	0.62	0.62	22	19	13.1	1.2
1.4	1.0	0.68	0.72	37	37	13.7	1.3
0.7	0.4	0.51	0.49	69	55	11.3	0.59
0.9	0.7	0.58	0.60	24	23	11.7	0.66

[b] Average of duplicate samples

parison among three Niskin bottles—one with rubber internal closures, a second with an internal Teflon-coated coil spring, and a newly designed "top drop" bottle with no internal closures. Samples were collected with these three samplers simultaneously at 10 m depth at a station several miles off the New Jersey and Long Island shores. Immediately upon recovery, duplicate aliquots of water were drawn from each of the bottles, filtered through a 0.4 μ Nuclepore filter, and acidified to pH 1 with silica-distilled nitric acid. The Niskin bottles containing the remainder of the samples were then allowed to stand for 3 hr in the ship's laboratory when a second set of duplicate aliquots was withdrawn, filtered, and acidified. The samples were analyzed for iron and zinc by direct injection flameless atomic absorption spectrophotometry (4). The results are presented in Table II. The analytical precision for these analyses is better than ±10% as determined from multiple analyses of replicate samples. As was expected from previous experience, the internal rubber closure bottle clearly contaminates the sample with zinc. Unexpectedly, however, it also con-

Table II.   Comparisons of Samples Collected by Three Designs of Niskin Bottle

	Top Drop	Silicone Rubber Spring	Teflon-Coated Coil Spring
Iron (μg/kg)[a]			
Drawn immediately	23	55	28
Drawn after 3 hr in bottle	20	33	23
Zinc (μg/kg)[a]			
Drawn immediately	1.3	6.2	1.6
Drawn after 3 hr in bottle	1.1	6.6	2.7

[a] Each number represents the average of multiple analyses (more than three) of each of duplicate samples.

taminates the sample with iron. Possibly, this iron is in the form of very fine particles which pass through the Nuclepore filter. This hypothesis is supported by the much larger drop in iron concentration with time in this than in both of the other bottles. This drop in "dissolved" iron concentrations is probably caused by agglomeration of colloids which become large enough to be retained by the filter. Newly formed colloidal-sized iron-containing particles generated from the rubber springs would be expected to agglomerate at a faster rate than the already aged material present naturally.

The use of a coil spring, whose new Teflon coat was not visibly damaged, contaminates the sample with small but significant amounts of both dissolved iron and zinc. Of particular interest, is the increase in zinc concentrations with time in all but the top drop bottle. Apparently, zinc continues to "dissolve" from the coil spring and possibly the rubber after closure, suggesting that deep ocean samples may be more contami-

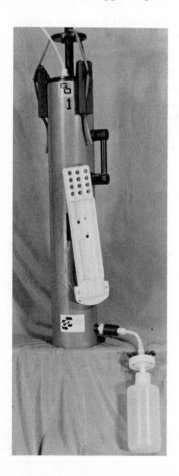

Figure 4. Top drop Niskin bottle with snap-on pressure fittings for filtration

nated than shallow samples because of the longer residence time of water in the bottle. The loss of zinc over the 3-hr period in the top drop bottle is small and only marginally significant compared with the analytical error. If the loss is real, then it may indicate that some zinc is associated with an actively precipitating phase such as that apparently removing the iron.

The top drop bottle is superior to other available bottles with respect to its lack of, or at least minimized, sample contamination for some trace metals. The top drop bottle has the additional advantage that it may be internally pressurized, permitting the seawater sample to be filtered without transfer to another container. This minimizes the risk of contamination of the water and particularly of the filter by the ship's atmosphere or other extraneous sources. A snap-on fitting has been developed so that a filter head containing a filter, preloaded in a dust-free atmosphere, may be attached to the petcock of the top drop bottle. A gas-tight nylon fitting is installed on the top drop in place of the air vent of normal Niskin samplers. A supply of filtered inert gas at excess pressures up to 1 atm is then coupled to this fitting and the water sample filtered without ever contacting the ambient atmosphere (Figure 4).

Top drop bottles were successfully and routinely operated on a Rosette system and on normal hydrowires during several cruises of the MESA New York Bight Project. Loading and handling of the bottles is minimally more difficult than for the older designs. They may, therefore, be used without difficulty in place of the older design for collection of samples for other than trace metal analysis. The use of such bottles for trace metal or suspended particulate chemistry would appear to be indicated.

Literature Cited

1. Brewer, P. G., Spencer, D. W., unpublished data.
2. Goldberg, E. D., "Marine Pollution Monitoring: Strategies for a National Program," p. 203, NOAA, Washington, D.C., 1972.
3. Robertson, D. E., *Anal. Chem.* (1968) **40**, 1067.
4. Segar, D. A., Cantillo, A. Y., ADVAN. CHEM. SER. (1975) **147**, 56.
5. Cooper, L. H. N., *J. Mar. Res.* (1958) **17**, 128.
6. Bader, H., University of Miami, personal communication, 1972.
7. Niskin, S. J., Segar, D. A., Betzer, P. R., in press.
8. Sigalove, J. J., Pearlman, M. D., *Undersea Technol.* (1972) **24**.
9. Segar, D. A., *Int. J. Environ. Anal. Chem.* (1973) **3**, 107.

RECEIVED January 13, 1975. This research was carried out as part of the National Oceanic and Atmospheric Administration (NOAA) Marine Ecosystem Analysis (MESA) New York Bight Project. The Environmental Research Laboratories of NOAA do not approve, recommend, or endorse any product, and the results reported in this document shall not be used in advertising or sales promotion or in any manner to indicate either implicit or explicit endorsement by the U. S. Government of any specific product or manufacturer.

3

Apparatus for the Sampling and Concentration of Trace Metals from Seawater

EARL W. DAVEY and ALBERT E. SOPER

National Marine Water Quality Laboratory, Environmental Protection Agency, South Ferry Rd., Narragansett, R. I. 02882

Apparatus and techniques for the sampling and concentration of particulate and dissolved trace metals (Cd, Cr, Cu, Fe, Mn, Ni, Pb, and Zn) from seawater were used to monitor ambient trace metal concentrations and heavy metal discharges. Particulate trace metals were collected in 0.4 μm Nuclepore filter bags, and dissolved metals were concentrated on Chelex-100 chelating resin. The apparatus is constructed of polyethylene and polypropylene plastics to reduce possible sample contamination.

The capabilities and limitations of analyses of water for trace metals have been described by Hume (*1*). From his discussion it can be inferred that few reliable methods have been available to analyze directly for particulate and dissolved trace metals in seawater. Therefore, it has been necessary to pre-concentrate seawater samples to reduce the high salt content and to increase trace metal concentrations to within the range of existing conventional instrumentation, such as flame atomic absorption spectrometry (AAS) or neutron activation analysis (NAA). Another problem, according to Robertson (*2*), is that seawater samples can be contaminated by trace metals during sampling and subsequent storage. Consequently, this paper describes apparatus which is designed to concentrate trace metals from seawater, thereby reducing some potential sources of sampling contamination errors.

Analysis Methods

In 1970 Davey *et al.* (*3*) reported on a modification of the technique of Riley and Taylor (*4*) which selectively removed trace metals from marine culture media. Radioactive trace elements were used to verify that the chelating resin, Chelex-100 (Bio-Rad, Richmond, Calif.), effectively removed certain trace elements from natural or artificial seawater (Table I). Essentially identical results were obtained in these experiments for either seawater collected from the Sargasso Sea or artificial seawater prepared according to Kester *et al.* (5). The water was radiolabeled with specific trace elements and passed through 10 ml sodium Chelex at 5 ml/min. However, trace metals were not completely released (3) from Chelex-100 by using the technique of Riley and Taylor (4). This problem was circumvented by the following modifications:

1. The ammonium rather than the sodium form of the resin is used as was later suggested by Riley and Taylor (6).

2. The volume of resin is reduced to 5 ml for a 4-l. seawater sample.

3. The flow rate is decreased to approximately 1 ml/min to compensate for the reduced resin volume.

4. Concentrated nitric acid is used for extraction and total release of the metals from the resin.

Table I.   Adsorption Efficiency of Purified Sodium Chelex-100 for Trace Metals in Seawater

Isotope	Adsorption (%)
^{65}Zn	>99
^{115m}Cd	>99
^{54}Mn	>99
^{64}Cu	>99
^{210}Pb	95
^{63}Ni	92
^{59}Fe	92
^{110m}Ag	33

The acid-extracted metals are usually analyzed by conventional flame AAS. However, the Chelex-100 can also be analyzed directly for cadmium, cobalt, manganese, and zinc by NAA if the resin is first rinsed with deionized water after sample passage to reduce salt content and then dried. The carefully purified ammonium form of Chelex-100 after NAA was essentially free from trace metal contamination.

Chelex-100 also competes against the added chelators glycine, histidine, and ethylenediaminetetraacetic acid (EDTA) (Table II). In these experiments, various concentrations of chelators or phytoplankton cultures with cells removed were mixed in seawater with ^{64}Cu, and the radioactivity was counted before and after the water was passed through Chelex columns. Since the only competition observed was with $10^{-6}M$ EDTA, the resin should be able to compete against natural complexes

Table II.   Competition for ^{64}Cu Between Chelex-100 and Various Chelators

Chelator	Concentration (M)	Competition (%)
Glycine	10^{-7}	0
	10^{-6}	0
	10^{-5}	0
Histidine	10^{-8}	0
	10^{-7}	0
	10^{-6}	0
EDTA	10^{-8}	0
	10^{-7}	0
	10^{-6}	34
Phytoplankton		0

having stability constants less than the $10^{-6}M$ EDTA–Cu complex in seawater.

Apparatus

The apparatus in Figure 1 is arranged to sample 0.1 m from the sediment–water interface when the polyethylene sampling tube is attached directly to the anchoring weight and 0.5 m below the air–water interface. However, intermediate depths could also be sampled by adjusting the bottom sampling tube to other desired depths.

The depth at which the water collection bottles are placed provides the hydrostatic driving force to fill the bottles with seawater through the connecting polyethylene tubing. The bottle is designed to fill with seawater from the bottom to the top to carefully exclude air. Thus, the collected seawater can be analyzed for oxygen, pH, salinity, and trace metals. The apparatus is constructed of polyethylene or polypropylene materials to reduce potential trace metal contamination to the sample. The only items not made from plastics are the weights.

The seawater from the sampler is poured into a 1- or 2-l. acid-cleaned and deionized water-rinsed linear polyethylene storage container. This can be converted later into a trace metal processing bottle by changing to a modified cap as shown in Figure 2. The processing bottles work in the following sequence.

The sample passes through a 0.4-μm Nuclepore (Pleasanton, Calif.) filter bag placed in a tubulated polyethylene container. The filter bags are made from 142-mm filters or, if a larger surface area is needed, by heat-sealing rectangles cut from 8×10-in. Nuclepore filter sheets. A small bead of silicone rubber is applied to the filter holder cap to make the filter assembly leak-tight. The first Chelex column removes the trace metals from the seawater. The second Chelex column is used as an analytical blank to correct for any salt matrix effects from the first column.

After collection, the samples are processed by the following method.

The water collection bottles are weighed before and after sample passage to determine the mass of water processed. The drained filter bags are removed from the plastic filter containers, placed into quartz vials, and digested with Ultrex (J. T. Baker Co., Phillipsburg, N.J.) concentrated nitric acid. The solution is taken to dryness, the residue is eluted with 0.5 ml of both Ultrex concentrated nitric and hydrochloric acids, and the eluate is diluted to 5 ml with deionized water. The Chelex material is removed from the columns, placed into quartz vials, rinsed with deionized water, and dried. The resin is digested in 3 ml Ultrex concentrated nitric acid for at least 1 hr before dilution to 10 ml with deionized water. The resin particles are removed by filtering through a porous plastic

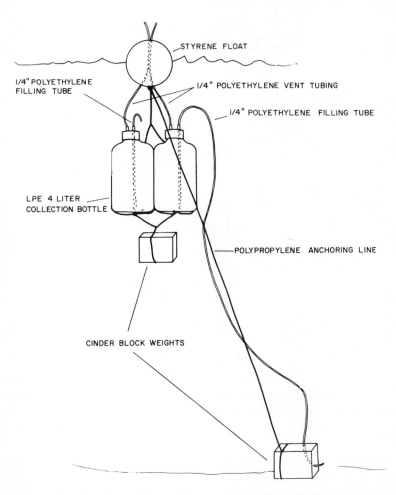

Figure 1. Apparatus for sampling water for trace metals analysis. It is designed to sample 0.1 m from the sediment–water interface and 0.5 m below the air–water interface.

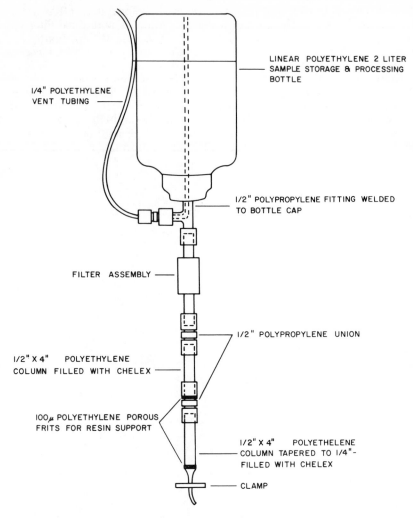

Figure 2.   Sample storage and processing bottle with modified cap, filter assembly, and columns

filter. Both the acid-extracted filters and resin are analyzed by AAS for Cd, Cr, Cu, Fe, Mn, Ni, Pb, and Zn.

Results

Table III presents results obtained from the sampling and concentration apparatus for both particulate and dissolved trace metals. The percentage of the ratio of dissolved-to-total trace metals also was computed. These results compare with values reported by others for Narragansett Bay (7).

Table III.   Concentration of Trace Metals from Narragansett Bay (25 October 1974)

Metal	Particulate	Dissolved[a]	Dissolved (%)
Cd	not detected	0.08 ± 0.03	100.0
Cr	0.30 ± 0.07	not detected	not detected
Cu	0.38 ± 0.07	0.74 ± 0.05	66
Fe	55.12 ± 3.39	0.30 ± 0.22	0.5
Mn	3.70 ± 0.32	11.36 ± 0.20	78.7
Ni	0.18 ± 0.06	4.08 ± 0.22	95.8
Pb	0.65 ± 0.07	0.51 ± 0.13	44.0
Zn	0.94 ± 0.26	5.28 ± 0.18	84.9

[a] All concentrations are expressed in μg/kg (ppb) as a mean ($\bar{x}$) of four replicates ($n = 4$) at the 95% confidence limit.

Discussion

The apparatus can theoretically sample to any depth. However, at depths greater than 100 m, it is logistically difficult to anchor and to retrieve. The apparatus could be used to monitor baseline trace metal levels, sewage and industrial metal outfalls, offshore dumping sites, and diffusion of heavy metals from polluted sediments. The concentration apparatus could also be adapted to monitor other compounds such as methylated mercury, chlorinated hydrocarbons, amino acids, etc. by replacing the Chelex-100 in the columns with other resins specific for the compounds to be monitored.

Literature Cited

1. Hume, D. N., ADVAN. CHEM. SER. (1967) **67**, 30.
2. Robertson, D. E., *Anal. Chem.* (1968) **40**, 1067.
3. Davey, E. W., Gentile, J. H., Erickson, S. J., Betzer, P., *Limnol. Oceanog.* (1970) **15**, 486.
4. Riley, J. P., Taylor, D., *Anal. Chim. Acta* (1968) **40**, 479.
5. Kester, D. R., Duedall, I. W., Connors, D. N., Pytkowicz, R. M., *Limnol. Oceanog.* (1967) **12**, 176.
6. Riley, J. P., Taylor, D., *Anal. Chim. Acta* (1968) **41**, 175.
7. Bender, M. L., University of Rhode Island, Graduate School of Oceanography, Narragansett, R. I., personal communication.

RECEIVED January 3, 1975

4

Considerations in the Design of *in Situ* Sampling Techniques for Trace Elements in Seawater

KENNETH E. PAULSEN, EDWARD F. SMITH, and HARRY B. MARK, JR.

Department of Chemistry, University of Cincinnati, Cincinnati, Ohio 45221 ‑

Accurate determination of trace element concentrations in natural water systems is becoming increasingly important. Present sampling and storage steps, however, cause tremendous changes in samples before they can be analyzed by more-than-adequate techniques. In order to avoid these alterations, we have suggested criteria for designing in situ sampling techniques to provide both quantitative and stable samples which can be directly analyzed with minimal handling and expense. The three methods which we have developed use electrodeposition, ion exchange membranes, and glass immobilized chelates. Of these, controlled pore glass immobilized chelates, such as 8-hydroxyquinoline, appear to be the most promising for proper sampling, as well as pre-concentrating, trace metals in natural waters.

One of the most important problems in oceanography and water resources science is the effect of the concentrations and concentration changes of trace metal ions on the nature of the water system (*1–6*). Recently, there has been much interest in the apparent increased concentration of metal ions such as mercury, lead, and iron. This concern is, at best, speculative since there are insufficient analytical techniques to establish baseline normal concentrations with the precision expected of good analytical methods. For example, there has been tremendous publicity concerning the level of mercury concentrations in edible fish in Lake St. Clair (*7, 8, 9*). Even in extreme cases, there was considerable disagreement in the true mercury concentrations in the fish analyzed. Rottschafer, Jones, and Mark (*9*) conducted a comparative study in which a homogenous sample of Coho salmon flesh was dis-

Table I.   Representative Results of Mercury Analyses in Edible Fish Tissue (*9*)

Method	Results (ppm)
Sample 1	
Atomic absorption [a]	0.60 ⎫
Atomic absorption [a]	1.0 ⎪
Atomic absorption [a]	0.85 ⎬ [b]
X-ray fluorescence	0.40 ⎪
Destructive NAA	0.87 ⎭
Ion exchange/NAA	0.86 (10% standard deviation)
Sample 2	
Ion exchange/NAA	0.58 (10% standard deviation)
NBS (non-des. NAA) [c]	0.54
Average results of 15 labs (various methods) [b]	0.51 (values ranged from 0.19 to 1.2)

[a] Atomic absorption results of three different laboratories.
[b] Results supplied by Grieg, Bureau of Comm. Fisheries, Ann Arbor, Mich.
[c] Results supplied by LaFleur of NBS (1970) (*10*).

Environmental Science and Technology

tributed to 15 laboratories for analysis for mercury. Numerous random samples of the homogenate were also analyzed by neutron activation by Dr. LaFleur (*10*) of NBS with the results showing less than 10% deviation. However, the overall results (Table I) showed an unacceptable spread. Similarly, a more extensive comparative study of trace elements in seawater by Brewer and Spencer (*11*) showed an even greater spread in results even though a carefully prepared sample which had no systematic error from sampling was used (Table II). Results for elements such as iron and cobalt, which analytical chemists feel are easily analyzed with accuracy, showed especially large variations. So the initial problem of studying trace metal concentrations is one of analytical chemistry.

Analytical methods have been developed which are sensitive enough to measure the low concentration levels of trace metals in seawater. Well defined methods, like emission spectroscopy, neutron activation analysis, anodic stripping voltammetry, atomic absorption spectroscopy, and mass spectroscopy, can be used individually or collectively to obtain the necessary data on trace metal concentrations. So why, even with these well developed methods, are we not getting reliable results from the analysis of trace metals in natural water?

Sampling Problems

The problem is not in the quantitative analytical techniques but in the methods used to obtain a representative sample quantitatively which is free from errors introduced during sampling and storage (*1, 2, 3, 12,*

13). In the past, the practice has been to take a sample from any depth in a large metal or (better) plastic container and then transfer the sample to another, usually plastic, container for subsequent analysis by appropriate analytical methods. Obviously, a metal container will contribute to the trace metal content of the sample, and even plastic containers will cause problems. Trace analysis studies have shown that plastic or glass sample containers can both absorb trace metal ions from the sample and/or contribute other metal ions to solution by surface dissolution (12, 13). Thus, the sample cannot be analyzed accurately because of the time-dependent effects on concentration which are related simply to the nature of the container and the conditions used to store the sample.

Table II. Elements Ranked in Order of Increasing Coefficient of Variation of the Grand Means, Calculated from the Pooled Standard Deviations (11)

Element	Coefficient of Variation	Abundance ($\mu g/kg$)
Sr	2.5	8100
F	2.8	1350
Rb	5.4	121
Cs	5.5	0.3
U	5.6	3.3
Sb	10.2	0.4
Zn	18.4	5
Cu	20.37	3
Co	22.0	0.1
Mn	24.5	1.5
Fe	25.8	14
Pb	29.2	5
Ni	33.5	2

Trace Element Intercalibration Study

One other problem also arises when the speciation of metal ions is of interest. The nature of the samples with respect to metal ion valence and/or degree of complexation can be expected to vary with changes in light intensity and energy, oxygen content, biological activity, pH, pressure, etc. Thus, the sample would best be maintained at the same dynamic equilibrium conditions from which it was taken, but this would be virtually impossible if a water sample were taken and stored for subsequent analysis.

At first glance, the only solution to this sampling problem is to make an *in situ* analysis without removing or perturbing the sample from its environment. *In situ* analysis has the further advantage of virtually instantaneous turnaround of results. Thus, if anything unusual is found at a particular location, the investigator knows about it immediately and can investigate it thoroughly. Several *in situ* analyses for selected species have been attempted (*14, 15, 16, 17, 18*). However, a system for complete *in situ* analysis presents many problems, especially expense (*17*), which make it unattractive at present for application to trace metal analysis.

Criteria for Developing in Situ Sampling Methods

Our approach to improving analytical results for trace metal analysis provides an alternative to the suggestion that the level of sensitivity and standardization of specific methods of analysis should be improved. We feel that, presently, there is a wide variety of suitable analytical methods if we can provide true and stable samples and minimize handling and manipulation. We have been developing *in situ* sampling methods to remove these sampling and human error factors and have been using the following design specifications and criteria.

1. The sytem must be sampled in a manner so that it is rendered physically and chemically inert.

2. The sample must be in a form to be analyzed by a direct non-destructive method—preferably with one step and no chemical preparative steps.

3. The analysis method must be one that is potentially applicable for shipboard or field use in order to get rapid analytical results.

4. The *in situ* instrumentation must be inexpensive, rugged, and durable.

5. The analytical method must provide high quality resolution for simultaneous identification of several elements.

6. The analytical method must also be quantitative with respect to initial concentration of each element in the original environment.

7. Samples must occupy a small space for convenient storage.
Three methods of *in situ* pre-concentrating/sampling have been or are being worked on in our laboratory which meet most of these design specifications.

Methods of in Situ Pre-concentrating/Sampling

Currently, work is being conducted on an *in situ* electrodeposition sampling device (*20, 21, 22*). It consists of a submersible, self-contained potentiostat, power supply, reference electrode, and working electrode. Metals are deposited on the 1-in. diameter, wax-impregnated, pyrolytic graphite working electrode (*21, 22, 23*) which can then be removed from the sampler at the surface and stored. The metal film can be either analyzed directly by x-ray fluorescence spectroscopy or emission spectroscopy, or it can be dissolved and analyzed by atomic absorption spectroscopy. The metals which can be sampled by electrodeposition are limited, however, to those with suitable reduction potentials.

We have done some experiments using ion exchange membranes (24, 25, 26), again looking for a simple direct method of obtaining a sample that is chemically and physically inert on removal from the measurement site and which can be directly (or simply) analyzed for trace element content. We used NAA of the membranes but found, however, that this membrane approach has limitations because of the long times required to reach the distribution equilibrium of even sufficient concentrations of trace elements in the membrane matrix. Other problems included interferences, such as the adsorption of organics, which affected the sensitivity and selectivity of the ion exchange (25).

Recently, Sugawara, Weetall, and Schucker (27, 28, 29) have developed a controlled pore glass (CPG) immobilized chelate which has several attractive properties as an *in situ* pre-concentrating/sampling matrix for trace elements. An amino-organosilane is covalently bonded to the surface of porous (550Å) glass particles (40–80 mesh and 80–120 mesh). A selected chelating group then reacts with the bonded organosilane, coupling the chelating group to the glass surface and rendering it immobile. The most successful chelate so far has been 8-hydroxyquinoline (Figure 1). The capacity of this material was determined in batch experiments and found to be in the range of 3 mg Cu^{2+}/ gm glass in a neutral pH solution (30). Again in batch experiments, the effect of pH on chelation of selected elements was determined by Sugawara *et al.* (29) and showed that, at neutral pH, elements such as copper, nickel, and cobalt are effectively chelated.

Using the glass immobilized chelate as a pre-concentrating/sampling matrix involves passing a constant flow of seawater (with a positive displacement Masterflex tubing pump, Cole-Palmer Instrument Co.) through a small column (2 mm × 15 cm) made in the form of a replaceable cartridge (Figure 2). This type of system is fairly simple, inexpensive, and rugged compared with adapting an analytical instrument for

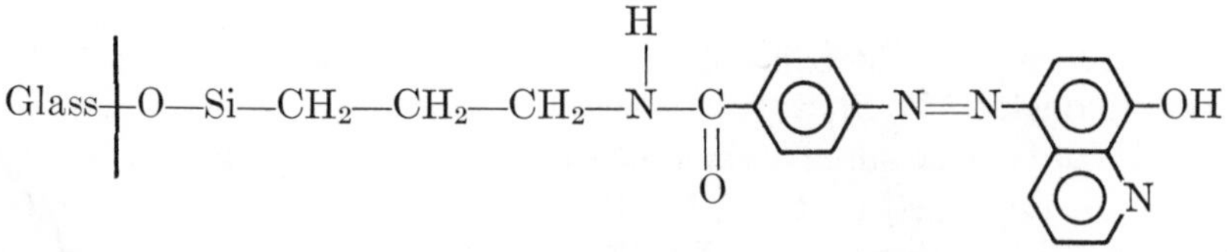

Figure 1.   Glass immobilized 8-hydroxyquinoline

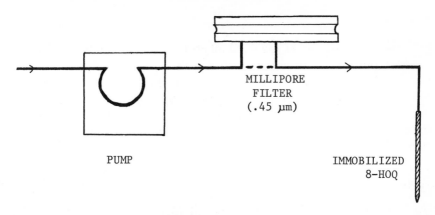

Figure 2. Flow system for seawater sampling/pre-concentrating on controlled pore glass immobilized 8-hydroxyquinoline (CPG-8HOQ)

in situ analysis. The forced water flow through the small mesh glass immobilized chelate will greatly increase the sensitivity compared with the membrane experiments. The glass immobilized chelate containing the trace elements can then be stored for subsequent analysis without significant loss of the trace elements. Analysis is by x-ray fluorescence, but NAA, atomic absorption, and emission spectroscopy can also be used.

An advantage of this sampling configuration is that particulate matter can be trapped on the column packing or removed by filtration prior to chelation. This allows analysis of either the total trace element concentration or the dissolved trace metal concentration.

Using the glass substrate is an improvement over organic ion exchange resins which have been used extensively by Riley (*31, 32*) and others (*33, 34*). The glass immobilized chelate does not expand or contract with loading or changes in ionic strength as do organic resins. Thus, the glass bead substrates give higher and more reproducible flow characteristics and, because the functional groups are distributed over the porous surface of the glass substrate, diffusion rates to the ion exchange sites are faster (*30*).

Another advantage of using the controlled pore glass matrix is that its selectivity can be changed by changing the chelating group. By altering the synthesis slightly, different chelating products have been made (*27*), and others are possible and are being investigated by ourselves and others (*36*). By synthesizing products specific for different groups of metal ions, it is possible that a mixed bed of such products would provide a matrix with a high capacity for a large variety of trace metals. Also, by using different immobilized chelating agents having different stability constants, information about the speciation of metal ions in seawater could be determined simultaneously with the analysis.

Conclusion

The development of these sampling techniques will thus solve the inherent problems encountered in the analysis of trace elements in seawater. The additional pre-concentration which occurs during sampling allows for the simple analysis of samples by existing techniques with even moderate detection limits. In this way, we will be better equipped to monitor the concentration of trace elements in seawater and determine their effect on the environment.

Acknowledgment

We wish to thank the University of Miami and J. S. Mattson for their valuable assistance in allowing us to use their facilities in field tests.

Literature Cited

1. Joe, J. H., Koch, H. J., "Trace Analysis," J. Wiley and Sons, New York, 1957.
2. Horne, R. A., "Marine Chemistry," p. 129, Intersciences, 1969.
3. Hume, D. N., ADVAN. CHEM. SER. (1967) 67, 30.
4. Martin, D. F., "Marine Chemistry," Vol. I and II, M. Dekker, New York, 1969 and 1970.
5. Joyner, T., Healy, M. L., Chakravarti, D., Koyanagi, T., Environ. Sci. Technol. (1967) 1, 417.
6. Goldberg, E. D., "Chemical Oceanography," J. P. Riley, G. Skirrow, Eds., Vol. I, Chapter 5, Academic, London, 1965.
7. Wallace, R. A., Fulkerson, W., Shults, W. D., Lyon, W. S., ORNL-NSF-EP-1, March, 1971.
8. Eyl, T. B., Wilcox, K. R., Jr., Reizen, M. S., Mich. Med. (1970), October, 873.
9. Rottschafer, J. M., Jones, J. D., Mark, H. B., Jr., Environ. Sci. Technol. (1971) 5, 336.
10. LaFleur, P., National Bureau of Standards, private communication, 1970.
11. Brewer, P. G., Spencer, D. W., "Trace Element Intercalibration Study," 70-62, Woods Hole Oceanographic Institute, 1970.
12. Robertson, David E., Anal. Chem. (1968) 40, 1067.
13. Robertson, David E., Anal. Chim. Acta (1968) 42, 533.
14. Riley, J. P., "Chemical Oceanography," J. P. Riley, G. Skirrow, Eds., Vol. II, Chapter 21, Academic, London, 1965.
15. Mancy, K. H., Okun, D. A., Reilley, C. N., J. Electroanal. Chem. (1962) 4, 65.
16. Albin, A. G., Ph.D. Thesis, Oregon State University, 1969.
17. Kester, D., Crocker, K., Miller, G., Jr., Deep Sea Res. (1973) 20, 409.
18. Whitfield, M., Limnol. Oceanogr. (1971) 16 (5), 829.
19. Mark, H. B., Jr., J. Pharm. de Belgique (1970) 25, 367.
20. Mark, H. B., Jr., Berlandi, F. J., Anal. Chem. (1964) 36, 2062.
21. Rottschafer, J. M., Ph.D. Thesis, University of Michigan, 1972.
22. Rottschafer, J. M., Boczkowski, R. J., Mark, H. B., Jr., Talanta (1972) 19, 163.
23. Vassos, B. H., Berlandi, B. J., Neal, T. E., Mark, H. B., Jr., Anal. Chem. (1965) 37, 1653.

24. Mark, H. B., Jr., Berlandi, F. J., Vassos, B. H., Neal, T. E., *Proc. Int. Conf. Mod. Trends Act. Anal., 1965*, (1966) 107.
25. Eisner, U., Rottschafer, J. M., Berlandi, F. J., Mark, H. B., Jr., *Anal. Chem.* (1967) **39**, 1466.
26. Eisner, U., Mark, H. B., Jr., *Talanta* (1969) **16**, 27.
27. Sugawara, K. F., Weetall, H. H., Schucker, G. D., 163rd Meeting Am. Chem. Soc., Boston, Mass., April, 1972.
28. Sugawara, K. F., Weetall, H. H., Schucker, G. D., private communications, 1972 and 1973.
29. Sugawara, K. F., Weetall, H. H., Schucker, G. D., *Anal. Chem.* (1974) **46**, 489.
30. Paulsen, K. E., Ph.D. Thesis, University of Cincinnati, 1975.
31. Riley, J. P., Taylor, D., *Anal. Chim. Acta* (1968) **40**, 479.
32. Matthews, A. D., Riley, J. P., *Anal. Chim. Acta* (1970) **51**, 287.
33. Nakagawa, H., Ward, F., *Public Health* Conf. Anal. Chem., Abs. (1960) 36.
34. Callahan, C. M., Rascual, J. N., Lai, Ming G., *U.S. C.F.S.T.I.* (1966) AD647661.
35. Biechler, D. G., *Anal. Chem.* (1955) **37**, 1055.
36. Hercules, D. F., *et al., Anal. Chem.* (1973) **45**, 1973.

RECEIVED November 27, 1974. This research was supported in part by the National Science Foundation, Grants NSF GP-33534 and NSF GP-35979X.

5

Ammonium Pyrrolidinecarbodithioate– Methyl Isobutyl Ketone Extraction System for Some Trace Metals in Seawater

RICHARD JAY STOLZBERG

New England Aquarium, Research Department,
Central Wharf, Boston, Mass. 02110

An evaluation of the factors affecting the reliability of the ammonium pyrrolidinecarbodithioate–methyl isobutyl ketone (APCD-MIBK) extraction system for Cd, Cu, Ni, Pb, and Zn in seawater established that pH control of the aqueous phase is important. The pH range 2.1 to 2.3 was the best for extracting these metals. In addition, the amount of APCD added was important both because of potential competition reactions and because of metal contamination in the APCD reagent. The overall precision of the technique and the precision of the atomic absorption measurements were calculated. This technique can be used effectively in both clean coastal waters and more polluted harbor waters.

Extractive pre-concentration techniques have been used for some time in trace metal analysis of seawater. In particular, the ammonium pyrrolidinecarbodithioate–methyl isobutyl ketone (APCD–MIBK) system has found moderate use (*1–7*). Because of the potential for multi-element analysis of a single extract and the applicability of the technique for shipboard use, this method was evaluated for the analysis of Cu, Cd, Pb, Ni, and Zn at natural levels in coastal waters. The results presented for those five metals show that this method is a reliable and valuable analytical tool for trace metal analysis in seawater if the limitations of the method are realized and the proper precautions are taken.

Procedure

The extraction procedures used in this study are shown in Figure 1. If total chromium is of interest, the right hand side is followed, the Cr (III) is oxidized to Cr(VI) with permanganate, and the resulting lower oxidation states of manganese are reoxidized to Mn(VII) with hydrogen peroxide prior to complex formation and extraction (6). In general, a 4-l. seawater sample is taken, filtered through an acid-soaked, deionized water-rinsed glass fiber filter if desired, and divided into four 800-ml subsamples in 1-l. glass separatory funnels. The subsamples are spiked with concentrated trace metal stock solution, the pH lowered to 2.2 with 6N HCl, the desired quantity of 5% APCD stock solution and 25 ml MIBK are added, and the extraction is carried out in a separatory funnel by manual shaking for 1 min. After 7 min, the phases are separated and the 11–13 ml of organic extract are stored in recappable polypropylene tubes. The extracts can be stored in a freezer for at least 24 hr with no change in calculated concentration for Cd, Cr, Cu, Pb, Ni, and Zn. If analysis for vanadium is to be performed, the analysis should not be delayed for more than a few hours (7).

Stock solutions of 5% APCD (Fisher Scientific) are made daily in distilled deionized water and extracted at least three times with MIBK. Seawater samples are stored in acid-soaked, deionized, water-rinsed polypropylene bottles. All reagents used were reagent grade or the highest grade readily available. Polypropylene graduated cylinders and separatory funnel stoppers and Teflon stopcocks were used in all cases.

Atomic absorption measurements were made using standard conditions. Nearly stoichiometric flames were used for all metals but chromium, for which a reducing flame was used. The air–acetylene flame was used for all metals but vanadium, for which a nitrous oxide–acetylene flame was used. A single slot titanium burner was used for all of the metals investigated. Water saturated MIBK was used as the blank. Table I presents typical instrument parameters.

Seawater samples were collected at two sites and were used to study various aspects of this extraction system. New England Aquarium "line water" is Boston Harbor water that has been allowed to settle and is then passed through a diatomaceous earth filter. Massachusetts Bay water was collected about 40 km east–northeast of Boston in an area typical of clean coastal waters (8).

Results

These studies demonstrate that at least three factors are quite important in determining the degree of reliability and accuracy of the APCD–MIBK extraction technique. The pH of the aqueous solution at the time of extraction controls the efficiency of extraction and thereby controls the precision, sensitivity, and, to some extent, the selectivity of the extraction. The other two factors, the presence of competing ligands and the quantity

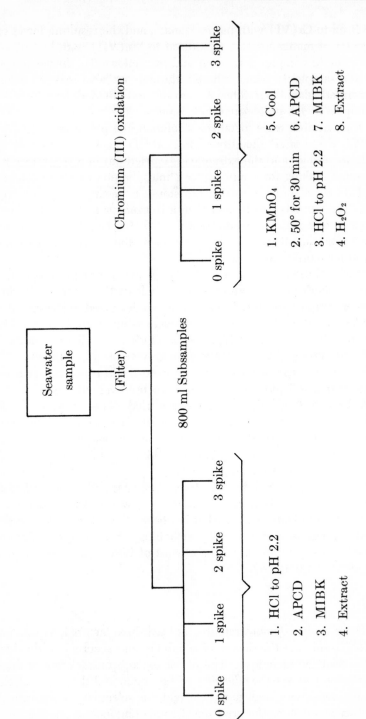

Figure 1.   APCD–MIBK extraction procedure

of APCD added, interact with one another and also control the extraction efficiency. In addition, the quantity of APCD added affects the accuracy of the technique because the APCD stock solution contains significant levels of some metals that are difficult to remove and thereby constitute a source of contamination.

Table I. Atomic Absorption Parameters[a]

Metal	λ (nm)	Flame	Burner Height (mm)	Lamp	Lamp Current (mA)
Cu	324.7	C_2H_2/air	12–18	Cu–Cr	4.5
Cd	228.8	C_2H_2/air	12–16	Cd	3.0
Cr	357.9	C_2H_2/air	~18	Cu–Cr	5.5
Pb	283.3	C_2H_2/air	12–14	Pb	3.5
Ni	232.0	C_2H_2/air	12–14	Ni	6.0
Zn	213.9	C_2H_2/air	12–18	Zn	4.5
V	318.4	C_2H_2/N_2O	~14	V	9.0

[a] Instrument, Instrumentation Lab model 153; solvent, methyl isobutyl ketone; aspiration rate, 3–4 ml min^{-1}; mode, 10 sec integration; slit width; 80 μ.

Influence of pH. Of all three factors, the pH of the aqueous solution is probably most important. There have been conflicting claims as to the pH range for effective extraction for some of the metals (*1, 2, 5, 9, 10*). The ranges quoted in the literature are considerably broader than those established in this work and extend, in general, to basic pH values where extraction of some of the trace metals studied here is much reduced. Whether the limited pH ranges for efficient extraction established here are a function of the low levels of metals we have been working with or a function of our matrix is not clear. Indeed, for Cd, Cr, Ni, Pb, and V at concentrations normally found in seawater, the pH of the aqueous phase is important. In general, in the range of pH from 1.5–8 the extraction is at a maximum from pH 2–2.5, decreases slowly through the pH range 3–4, and drops finally to much reduced values in slightly basic solution. Vanadium and chromium are especially sensitive to pH, and little extraction is noted if the pH is greater than 4 or 5 (*6, 7*). As an example, Figure 2 demonstrates the drop in Ni–PCD extraction as the pH of the extracted aqueous solution varies over the range 1.5–7.5. The extraction of copper and zinc as PCD complexes depends much less on pH as shown in Figure 3 for the copper complex. Here the decrease in extraction efficiency is a great deal smaller over the same pH range. A pH plateau exists in the range 2.0–2.5 for all seven metals tested where extraction efficiency is both high and constant. By attaining a pH of 2.2 ± 0.1 after extraction, it is possible to get quite reproducible sensitivities from sample to sample.

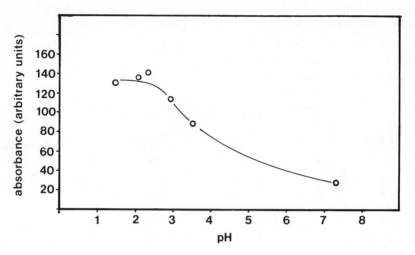

Figure 2.   pH Dependence of Ni–PCD extraction from line water

Effects of Added Ligands. Investigation of the effects of ligands added to seawater samples was undertaken to determine whether the presence of these compounds might interfere with the formation and extraction of metal-PCD complexes. Glycine, thiamine, citrate, cysteine, EDTA, salicylate, pyrogallol, and phosphate were the ligands studied.

Typically a 4-1. sample of line water was spiked to approximately 100 ppm with a single ligand, the sample was split into four 800-ml portions, spiked with metal standard, and extractions were carried out using 4 ml of 5% APCD (200 mg) per 800 ml subsample. No significant

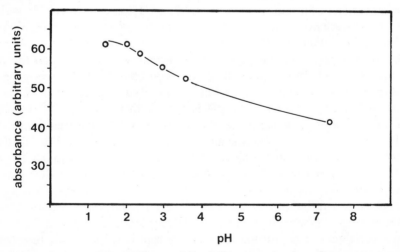

Figure 3.   pH Dependence of Cu–PCD extraction from line water

changes in the slope of the calibration curves obtained were observed except in the cases of nickel and zinc in the presence of EDTA. In these two cases, significantly lower calibration curve slopes were observed, indicating competition between the PCD moiety and the added EDTA. Table II presents some of the results of more detailed studies with EDTA as the competing ligand. A significant decrease in calibration curve slopes occurs for zinc and nickel with an increase in EDTA concentration. No significant differences in calibration curve slope were noted for lead and cadmium. The increases in average calibration curve slopes shown in Table II for zinc, nickel, and copper as the concentration of EDTA is increased from 0 to 10 ppm are not significant in this particular experiment and have been observed to be significant only for copper in two similar experiments. The reason for this apparent increase in copper calibration curve slope with the addition of a small amount of EDTA is not clear.

Table II.   Effect of EDTA Concentration on Calibration Curve Slope[a]

Na_2 EDTA Added to Line Water (ppm)	Zn Slope	Ni Slope	Cu Slope
0	9.5(1.1)	16.6(5.7)	38.8(9.8)
10	12.3(1.7)	21.8(4.9)	57.1(14.1)
100	8.0(0.4)	9.6(3.7)	41.8(5.0)
500	3.9(0.3)	1.8(0.1)	49.7(0.4)

[a] Each value is the average slope (and standard deviation) for two extractions of line water. 200 mg APCD added to 800 ml line water.

Effects of Variation of Quantity of APCD Added. As originally developed in our laboratory, this method was used to analyze trace metal levels in Boston Harbor water. It was established that 200 mg of APCD for 800 ml water was the minimum necessary to ensure a high degree of reliability in the extraction step, presumably caused by the elevated levels of organic and particulate matter. However, since coastal water typically contains much lower levels of organic and particulate matter than does Boston Harbor water, it seemed that a systematic investigation of the effects of lowering the quantity of APCD added might be useful. The amount of APCD added to 800 ml line water could be reduced from 200 to 10 mg with only a slight decrease in calibration curve slope for Cu, Cd, Ni, and Pb. A rapid decrease in slope was observed for zinc if the amount of APCD added was reduced to below 50 mg. Figure 4 demonstrates the different behavior of copper and zinc in this respect. The error bars for copper represent the standard deviation of the slope as calculated in the linear regression analysis of the data from one extraction series.

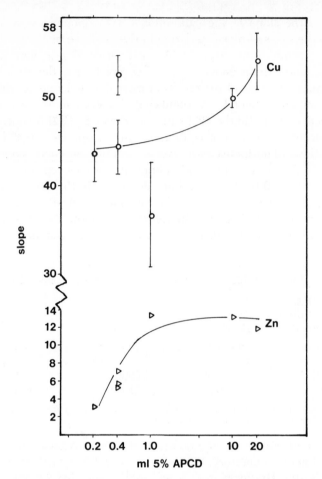

*Figure 4. Variation of calibration curve slope with
volume of 5% APCD added to 800 ml line water*

Another effect of reducing the quantity of APCD added is that the
effect of EDTA on the calibration curve slope is magnified. Comparison
of Table III with Table II shows that the effect of 100 ppm EDTA on
the calibration curve slopes for zinc and nickel is much more pronounced
if only 20 mg APCD is added instead of 200 mg. If one assumes a simple
competition between EDTA and pyrrolidinecarbodithioate, this behavior
is not at all surprising.

Metal Contamination by APCD Stock Solution. During the course
of this work suspicion arose that an appreciable quantity of some transition metals remained in the 5% APCD stock solutions even after repeated
extractions with MIBK. Verification of this fact was obtained in three
independent observations. The results of the three independent ap-

Table III. Effect of 100 ppm EDTA on Calibration Curve Slope with 20 mg APCD Added per 800 ml Water

Na_2 EDTA Added to Line Water (ppm)	Zn Slope (and Standard Deviation)	Ni Slope (and Standard Deviation)	Cu Slope (and Standard Deviation)
0	5.8(1.2)	16.8(7.0)	47.5(4.4)
100	1.0	1.6	55.9

proaches closely agreed with one another as to the expected levels of contamination caused by unextractable quantities of cadmium, nickel, and lead in the stock APCD solutions.

Poor agreement existed in the trace metal data from 24 sites in the Massachusetts Bay Foul Area obtained by both this extraction–AA method and by differential pulse anodic stripping voltammetry (8). The individual site differences in metal concentration as determined by the two methods were calculated as was the t-value for each metal. The data in Table IV demonstrate a significantly larger concentration of cadmium and lead determined by the AA method as compared with the DPASV method. The data for zinc show a small but insignificant difference in the average concentrations, indicating no visible bias. The case of copper, where the difference is significant but the direction of the difference is opposite that of lead and cadmium, appears to arise from electrochemical effects and will not be considered in this paper.

Although increasing the quantity of APCD added in the extraction changed the slope of the calibration curves but little, it had a profound effect on the calculated concentration of nickel, lead, and cadmium. In Figure 5a, a definite increase is observed in the concentration of cadmium, lead, and nickel as the quantity of 5% APCD added per 800 ml was increased beyond approximately 1 ml. Figure 5b presents the results for copper and zinc. Copper is unaffected by the quantity of APCD added. The case for zinc is less clear, no doubt because of the decrease in calibration curve slope and the resulting decrease in precision as the quantity of APCD is reduced below 1 ml per 800 ml of sample. The

Table IV. Comparison of AA and DPASV Data for Massachusetts Bay Water[a]

Metal	Average d	Calculated t	df	t(95%)
Cd	−0.16	5.67	23	2.07
Cu	0.54	1.96	19	2.09
Pb	−0.88	5.43	23	2.07
Zn	0.22	0.30	18	2.10

[a] d = ppb metal determined by ASV minus ppb metal determined by AA; 4 ml 5% APDC added to 800 ml seawater. $t = \dfrac{d}{S_{\bar{d}}}$; $S_{\bar{d}}$ = standard deviation of the average d.

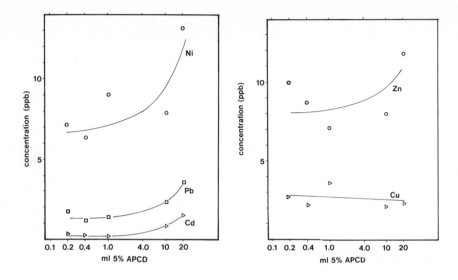

*Figure 5.   Calculated metal concentration as a function of volume of 5%
APCD added to 800 ml line water. Left (a): Cd, Ni, Pb. Right (b): Cu, Zn.*

lines drawn in Figures 4 and 5b were calculated as a linear least squares
fit to the data. This line was then plotted in the more convenient semi-
logarithmic form.

Since these two observations pointed to metal contamination caused
by the APCD stock solution, both extracted and unextracted APCD
solutions were analyzed for metal content.

Two bottles from two different lots of APCD were analyzed in duplicate. From each bottle, 100 ml of 5% solution was made, and two 25 ml
portions were transferred into round bottom flasks. The remaining 50 ml

		Table V. Metal Content of Unextracted	
Bottle	*Lot*	*Cd*	
1 Unextracted	A	0.8(0.3)	
1 Extracted	A	1.0(0.4)	
2 Unextracted	A	0.8(0.1)	
2 Extracted	A	0.5(0.1)	
3 Unextracted	B	0.7(0.1)	
3 Extracted	B	0.5(0.1)	
4 Unextracted	B	0.6(0.1)	
4 Extracted	B	0.4(0.2)	
Average unextracted		0.7(0.2)	
Average extracted		0.6(0.3)	

[a] Each value is the average concentration (and standard deviation) calculated from

was extracted with 30 ml MIBK and then twice more with 20 ml MIBK. The extracted solution was then divided into two equal portions and transferred to round-bottomed flasks. The volume of each sample was reduced to approximately 5 ml. Nitric acid and sulfuric acid were added, and the concentrate was digested under reflux for 30 minutes. Suitable blanks and spiked samples were also carried through the procedure. The digestates were diluted to 25.0 ml and were analyzed by AA spectrophotometry.

The results are presented in Table V. The presence of $\mu g/g$ quantities (dry weight basis) of Cd, Ni, Pb, and Zn in the APCD even after MIBK extraction represents a very real source of contamination when examining ng/g quantities in seawater. Copper levels were indistinguishable from zero.

The magnitude of the trace metal contamination from the addition of 200 mg APCD per 800 ml seawater was calculated for all three sets of observations, and the results are presented in Table VI. The agreement between the sets of numbers is quite good, and the contamination expected in the case of cadmium, nickel, and lead is quite significant compared with metal levels found even in harbor waters.

Precision of the Technique

The overall precision of the APCD–MIBK extraction technique coupled with atomic absorption analysis was obtained by analysis of data obtained from replicate analysis of line water over a period of a few months. For each experiment performed, two, three, or more extractions of line water were always performed in the normal fashion (4 ml 5% APCD added). Thus, although the trace metal composition of the line water changed over this period of time, an estimate of the precision of

and Extrated APCD (μg metal/g)[a]

Ni	Cu	Pb	Zn
1.8(0.8)	0.7(0.5)	3.6(1.1)	0.6(0.1)
2.4(1.9)	−0.1(0.2)	1.5(1.9)	0.7(0.7)
4.6(0.5)	0.5(0.2)	3.4(1.2)	1.0(0.8)
4.6(2.4)	—	4.1(0.5)	0.5(—)
3.4(2.6)	0.3(0.1)	6.5(2.1)	0.8(0.0)
2.2(2.7)	0.0(0.0)	2.7(3.1)	0.2(0.3)
2.7(1.0)	0.2(0.2)	4.7(—)	0.9(0.5)
2.2(1.8)	−0.1(0.2)	3.0(0.8)	0.9(0.1)
3.1(1.6)	0.4(0.3)	4.5(1.8)	0.8(0.4)
2.8(2.0)	−0.1(0.2)	2.8(1.7)	0.6(0.4)

one absorbance measurement on two digestates from the same original APCD solution.

Table VI.   Magnitude of Trace Metal Contamination Caused by APCD[a]

	Zn	Cu	Cd	Ni	Pb
Calculated contamination determined by acid digestion of APCD (ppb)	0.1	0.0	0.15	0.7	0.7
Calculated by varying volume of APCD added to sample (ppb)	0.5	−0.1	0.24	1.1	0.5
Observed difference between AA and ASV data for Mass. Bay water (ppb)	−0.1	−0.5	0.16	—	0.8

[a] 4 ml of 5% APCD was used in 800 ml seawater.

the technique for each metal could still be made using these data. The standard deviation of the average metal concentration for each experiment was calculated. The variance (standard deviation squared) was multiplied by the number of degrees of freedom for the particular experiment to give a weighted variance (or a sum of the squares of the standard deviation). The sum of the weighted variances was calculated and divided by the total number of degrees of freedom to obtain the variance of the entire technique. The calculated values of the standard deviation for each metal concentration, *i.e.*, the square root of the variance, are presented in Table VII.

The contribution of the variance of the atomic absorption analysis to the total variance was calculated from duplicate AA analyses of the same extract, usually at the beginning and end of a series of AA measurements of a single metal. The variance of the AA method and the variance of all other factors, including the extraction itself, were calculated and are presented in Table VIII. It can be observed that the contribution to the total variance by the AA measurement is substantial for Cd, Ni, Pb, and Zn. The relative standard deviation from other sources, including the extraction technique itself, contamination, and variation between sup-

Table VII.   Overall Precision of APCD–MIBK Trace Metal Extraction Scheme with Atomic Absorption Analysis

Metal	Nominal Line Water Concentration (ppb)	Standard Deviation of Technique	Relative Standard Deviation (%)	Degrees of Freedom
Cd	0.5	0.15	30	22
Cu	1.4	0.34	24	25
Ni	4.3	1.17	27	9
Pb	2.0	0.61	31	16
Zn	7.4	0.94	13	20
V	3.2	0.7	22	12

posedly identical samples drawn at the same time, is quite good for nickel and zinc. This figure of 7% probably represents the upper limit of the relative standard deviation of the extraction technique for these metals. It is possible that the relative standard deviation of the extraction technique itself for cadmium is considerably lower than 16%. This inflated value could be ascribed to variations in contamination arising from the ubiquitous cadmium associated with urban laboratories. The reasons for the poor precision for extraction of copper and lead have not been elucidated.

Table VIII.   Sources of Variance in Calculated Concentration of Trace Metals Using APCD–AA Technique

Metal	*Cu*	*Pb*	*Zn*	*Cd*	*Ni*
Nominal concentration (ppb)	1.6	1.9	8.0	0.45	4.5
Total variance	0.116	0.37	0.88	0.022	1.37
Variance of AA	0.017	0.16	0.58	0.017	1.28
Variance from other sources, including extraction technique by difference	0.099	0.21	0.30	0.005	0.09
RSD of other sources, including extraction technique (%)	20	24	7	16	7

Discussion and Conclusions

A comparison of our results with those of others investigating this same technique shows significant differences between what have been observed to be important factors. The role of metal contamination from the APCD has not been extensively studied previously. Of the papers cited (1–7), only one (5) touches directly on that point. They note that after carefully purifying the APCD in an unspecified manner, trace metal contamination was not observed. However, they did note a constant AA blank which they ascribed to water and inorganic salts in the MIBK phase. They made this statement on the basis of correlating the blank signal absorbance with the wavelength at which they were performing the measurements. It should be noted, however, that their data also correlate well with values of solubility products for the various metal sulfides (except for iron). It is also possible that the purity of the reagent varies over a wide range of values.

While it has been reported that some metal–PCD complexes are efficiently extracted over only a narrow range of pH values (2, 6, 10), two reports that are quite analogous to this one find excellent extraction over a wide pH range. Brewer *et al.* (5) find greater than 80% extraction of Zn, Cu, Ni, Co, Fe, and Pb from seawater over the pH range 2–7. They calculated this by the ratio of the first and second extractions on the same seawater sample. Brooks *et al.* (*1*) found that the extractions worked equally well at pH 8 and in the pH range 4–5 for the same six metals. These results are contrary to what we observed.

The precision of the technique for seawater analysis as presented in the literature (*1, 5*) tends to be considerably better than we have observed here. The values obtained in other papers were for duplicate analysis of the same sample and were most likely extracted sequentially from the same bulk sample and analyzed one directly after the other. This was not the case here because the data analyzed in this paper were not generated specifically to analyze the ultimate precision of the technique. Line water samples run normally were as a rule interspersed throughout the test samples. A number of water samples would be drawn at the start of an experiment and stored unacidified in 4-l. polypropylene bottles. Over the course of up to 6 or 8 hr, extractions would be performed so that differences in trace metal concentration might be expected between replicates run early and late in the experiments. This factor, which allows for significant adsorption and/or desorption of trace components, could readily explain our high standard deviations. We feel that this approach is valid to determine the precision of the technique in the field where non-optimum conditions often occur and where the factor of time between sampling and analysis is often an uncontrollable variable. It is likely that the actual precision of this technique in the field lies between those values calculated here and elsewhere (*1, 5*).

Comparison of this technique with others is difficult. Neutron activation analysis and isotope dilution mass spectrometry are used for the analysis of trace metals in seawater but are in a class by themselves in terms of equipment needed. Co-precipitation techniques are generally applicable to only a limited number of metals per precipitating agent. A technique that does compete directly with the APCD–MIBK/AA technique is that of pre-concentration of trace metals on chelating ion exchange resin followed by AA analysis of the acid effluent (*11, 12*). While the efficiency of uptake of added metal is excellent, there is some doubt about the ability of Chelex 100 to concentrate quantitatively both bound and labile species of Cu, Pb, Cd, and Zn (*13*). Indeed, this is a problem that has not been investigated for the APCD extraction technique. The choice between the ion exchange pre-concentration technique and the APCD pre-concentration technique is likely a matter of a large number of factors including time between sampling and ultimate analysis, number of samples to be run, presence of organic and particulate matter, identity of the metals of interest, concentration of metals, and equipment available.

It should be stressed that the APCD–MIBK extraction technique is reliable for routine analysis of coastal waters with reasonable precision considering the nature of the problem to be solved. However, one must be aware of the contamination problem inherent in the system. As a result of the poor extraction of cadmium–, nickel–, and lead–PCD complexes at the pH of a 5% APCD solution (pH 8.2), removal of these metals by the simple extraction with MIBK is not feasible. However, there is a viable alternative to painstakingly preparing APCD solutions that are metal free. The quantity of APCD added to a seawater sample can be reduced to levels where the contamination problem is very small. Only a slight decrease in sensitivity is realized for Cd, Cu, Ni, and Pb. In the case of zinc, although a large decrease in extraction efficiency and sensitivity is realized, the relatively large concentrations of this metal found in coastal waters and high sensitivity of the AA technique would probably allow for precise determinations. In the case of analyzing more polluted water where larger quantities of APCD are necessary to ensure reliable extraction, suitable blank corrections must be made.

Acknowledgment

I wish to thank Dennis Carson for his assistance in performing part of the experimental work.

Literature Cited

1. Brooks, R. R., Presley, B. J., Kaplan, I. R., *Talanta* (1967) **14**, 809.
2. Fishman, M. J., Midgett, M. R., "Trace Inorganics in Water," ADVAN. CHEM. SER. (1968) **73**, 230.
3. Morris, A. W., *Anal. Chim. Acta* (1968) **42**, 397.
4. Mulford, C. E., *At. Absorpt. Newsl.* (1966) **5**, 88.
5. Brewer, P. G., Spencer, D. W., Smith, C. L., "Determination of Trace Metals in Sea Water by Atomic Absorption Spectroscopy," Atomic Absorption Spectroscopy (1969) *ATSM Tech. Publ.* 443, 70.
6. Gilbert, T. R., Clay, A. M., *Anal. Chim. Acta* (1973) **67**, 289.
7. Ladd, K. V., Ph.D. Thesis, Brandeis University, 1974.
8. Gilbert, T. R., "Studies of the Massachusetts Bay Foul Area," New England Aquarium Report No. 1–75, The Commonwealth of Massachusetts Division of Water Pollution Control, 1975.
9. Lakanen, E., *At. Absorpt. Newsl.* (1966) **5**, 17.
10. Christian, G., *Anal. Chem.* (1969) **41**, 24A.
11. Goya, H. A., Lai, M. G., "Adsorption of Trace Elements from Sea Water by Chelex 100," U.S. Naval Radiological Defense Lab., TR-67-129, 1969.
12. Riley, J. P., Taylor, D., *Anal. Chim. Acta* (1968) **40**, 479.
13. Florence, T. M., Batley, G. E., *Talanta* (1975) **22**, 201.

RECEIVED January 3, 1975. This work was supported in part by Project No. 74-06, Department of Natural Resources, Commonwealth of Massachusetts.

6

Determination of Trace Metals in Aqueous Solution by APDC Chelate Co-precipitation

EDWARD A. BOYLE and JOHN M. EDMOND

Department of Earth and Planetary Sciences, Massachusetts Institute of Technology, Cambridge, Mass. 02139

A rapid technique has been developed for quantitatively concentrating several trace metals from aqueous solution. The metals are co-precipitated as dithiocarbamate chelates by adding an excess of another dissolved metal. This technique has been coupled with atomic absorption analysis for the precise determination of nmol/kg quantities of copper in seawater. Radiotracer experiments show that nickel, iron, and cadmium are also co-precipitated by this technique under proper experimental conditions.

The precise determination of transition metals in natural waters is difficult because the levels encountered challenge the detection limits of the available instrumentation. Even when the sensitivity of a method is sufficient, the background of other constituents, such as sea salt, often masks the detection of an element through secondary effects, such as broadband absorption in flameless atomic absorption or background radiation in neutron activation. Chemical separation techniques have been used to overcome these problems, both to increase the concentration above detection limits and to remove interfering constituents. Solvent extraction has proved useful in many instances, but care must be taken to ensure phase equilibration, and large-volume sample extractions are not practicable. Multistage or continuous solvent extraction techniques have been suggested but have not found widespread applicability. Large-volume, single-stage separation has been achieved through the methods of co-precipitation and co-crystallization, but the methods are time consuming and the recoveries sufficiently irreproducible that yield monitors are necessary. These difficulties have discouraged widespread use of such methods. A technique which combines rapidity and reproducibility with large-volume extraction capacity is clearly desirable.

Dithiocarbamates (RCS_2^-) have been applied widely in the analytical chemistry of trace metals in natural water (*1, 2, 3*). The ammonium salt of the dithiocarbamate of pyrrolidine (APDC) is a water-soluble compound which forms water-insoluble uncharged chelates with a variety of metals. At the very low metal concentration levels found in natural waters, dithiocarbamate chelates form colloids (*4*). At higher concentration levels the chelates precipitate (*5*). Since the metal ion in the chelate is effectively shielded by hydrophobic groups, precipitation in a multi-metal system is relatively nonselective; the chelates of several metals precipitate together rather than as separate phases. If one transition metal is present in sufficient concentration to form particles, other metals will be incorporated into the particles regardless of their concentration. The analytical implication of this behavior is clear—APDC chelates of trace transition elements can be co-precipitated by addition of one transition element up to sufficient concentration to form chelate particles.

On the basis of the above phenomenon, we developed a rapid quantitative separation technique capable of concentrating several trace transition metals from large-volume samples. Addition of Co^{2+} to the samples produces filterable particles upon mixing in APDC. The solution is then filtered and the precipitate brought into solution for analysis. By use of this technique, large volumes of dilute sample can be quickly reduced to a much smaller volume of concentrated solution for analysis. The sample size is only limited by that of workable container and filter sizes. We have routinely concentrated from 4 l. within 30 min using a 47-mm diameter glass fiber filter; much larger volumes could be handled easily using larger filters. The procedure has a high yield (> 90%) for several metals and is quite reproducible. We have coupled this concentration technique with atomic absorption spectroscopy in the precise analysis of copper at the nmol/kg level in seawater.

Basic Procedure

Reagents

1. Hydrochloric acid: 6N hydrochloric acid is redistilled in a Vycor still.

2. APDC: synthesized (*6*) and recrystallized once from ethanol.

3. APDC solution: 10 g APDC are dissolved in 500 ml of distilled water, and the solution is purified by repeated solvent extraction with carbon tetrachloride.

4. Buffer: 170 g ammonium acetate is dissolved in 500 ml of distilled water. 5 ml of APDC solution is added, and the solution is purified by repeated solvent extraction with carbon tetrachloride.

5. Cobalt solution: 0.4 g of reagent grade $CoCl_2 \cdot 6H_2O$ is dissolved in 500 ml of distilled water. This solution contains 2000 μg of Co^{2+} per 10 ml.

6. MEK–HNO$_3$: 50 ml of 0.1N nitric acid (prepared from distilled water and reagent grade nitric acid) is dissolved in 1 l. of redistilled methyl ethyl ketone.

7. Acetone–HNO$_3$: 100 ml of 0.1N nitric acid is dissolved in 1 l. of redistilled acetone.

Pre-concentration

1. Acidification: 5 ml of redistilled 6N hydrochloric acid is added for each liter of sample (added immediately upon collection to stabilize the solution from adsorption and precipitation).

2. Cobalt addition: 2.5 ml of cobalt solution is added per liter of sample.

3. Buffering: 10 ml of buffer solution is added per liter of sample, and the solution is well mixed. The pH of the solution should then be near 4.

4. APDC addition: 5 ml of APDC solution is added per liter of sample, and the solution is well mixed. Slow stirring may be continued to produce larger particles.

5. Filtration: The solution is vacuum-filtered through an acid-leached glass fiber filter and rinsed with 2 ml of distilled water.

6. Drying: The filters are dried for a few hours at 60°C.

Atomic Absorption Determination. The precipitate is washed through the filter into a 25-ml filter flask with 20 ml of MEK–HNO$_3$. This solution is then aspirated into the atomic absorption burner (we used a

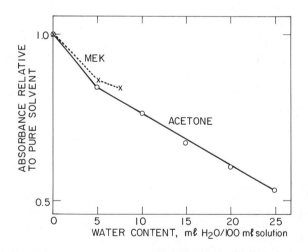

Figure 1. AA solvent enhancement for copper as a function of water content for methyl ethyl ketone and acetone

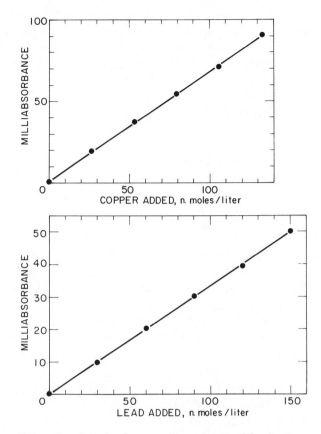

Figure 2. Standard curves for copper and lead using method described in text and flame atomic absorption. Copper was determined at 324.7 nm, and lead was determined at 283.3 nm.

Perkin Elmer model 403) using the pure solvent as a baseline. Standards made up in the pure solvent are used for calibration, and the method of standard additions is used to monitor recovery efficiency.

Comments on the Procedure

The amounts and proportions of the various reagents are optimized to give complete recovery with least reagent addition so as to minimize blanks. The APDC is in large excess relative to cobalt to avoid competition of the trace metals with cobalt for the chelating agent. If the particles produced are not removed by the filter used, the cobalt and APDC should both be increased in the same proportion until the filtration completely removes the chelates.

The drying step is necessary because the organic solvent enhancement in flame atomic absorption depends on the water content of the solvent. Figure 1 shows the relative variation in solvent enhancement caused by changes in water content for methyl ethyl ketone and for acetone. If the filters are not dried, the moisture on the filter increases the water content of the solvent and lowers the absorbance, so that erroneously low recovery efficiencies are calculated.

Cobalt is used as the precipitant because its APDC chelate is readily soluble in organic solvents. If the cobalt chloride produces high blank values, it may be purified by anion exchange (7). Other metals have been used, but not all are readily soluble. The copper chelate is readily soluble whereas the lead and nickel chelates are not.

Experimental Results

Copper and Lead. Using the conditions described above, copper and lead determinations were made on 1-l. distilled water solutions spiked with a known quantity of copper and lead. The results of these analyses are shown in Figure 2. Both copper and lead give linear standard curves. For copper the recovery is 97%, and the standard error is ±0.5 milliabsorbance units. For lead the recovery is 95%, and the standard error is ±0.4 milliabsorbance units. Both copper and lead are recovered reproducibly with high yields.

Copper in Seawater. Copper occurs in open-ocean seawater at very low levels, a few nanomoles per kilogram. Accordingly, the precipitation procedure was scaled up to use 4-l. samples. The above procedure was

Table I.   Copper Spike Experiments

Sample Depth	Unspiked (nmol/kg)	Spike (nmol/kg)	Spiked Total (nmol/kg)
398	5.94	14.77	18.97
1106	3.44	14.77	16.31
2189	5.87	14.77	19.17
300	2.21	23.42	22.51
635	2.78	23.42	23.85
906	2.90	23.42	23.82
501	3.62	25.09	25.56
1306	3.52	25.09	25.58
1507	3.60	25.09	26.18
61 F[a]	6.33	7.32	12.99
1306 F[a]	4.76	7.32	11.16
2544 F[a]	6.34	7.32	12.74
2581 F[a]	5.27	7.32	11.89

[a] F denotes filtered sample. Groupings indicate samples run on the same day.

followed except that the filters were not dried (the solvent enhancement effect was not appreciated at that time), and acetone–HNO_3 was used as the solvent system. The determinations were carried out on acidified seawater samples taken from 30-1. PVC Niskin bottles from a profile at 21°N 109°W over the East Pacific rise. Each sample was determined once as is and once with a known spike of copper added. The recovery efficiency was computed from comparison of the spike absorbance with standards in the same solvent system. The recovery efficiency as calculated in this manner is a few percent lower than the true recovery efficiency because of the solvent enhancement effect mentioned previously. The results of this experiment are listed in Table I, and the complete profile for unfiltered samples is shown in Figure 3. The recovery efficiency was constant within 2% between spikes run on the same day and between days—this indicates the precision of the analysis. The inherent precision of this method of copper determination in seawater is on the order of ±0.2 nmol/kg. While the levels found are much lower than previously reported, the scatter in the data is also large, and filtered samples are higher than unfiltered samples. Since we know of no mechanism to maintain such distribution in the water column, we suspect contamination of the samples during collection. This problem appears to be the limiting factor in trace metal analyses in seawater at the present time. When extreme care is taken to avoid contamination, oceanographically consistent results are obtained (8).

On some of the samples, lead determinations were also made; however, lead was near or below the detection limit value (~ 0.5 nmol/kg). Consequently, this method is not suitable for the analysis of lead in

on Seawater Samples

Blank (nmol/kg)	% Recovery	Daily Av. % Recovery
.50	88.2	88.4 ± 1.6
.50	87.1	
.50	90.0	
.73	86.7	
.73	90.0	88.7 ± 2.0
.73	89.3	
.29	87.5	
.29	87.9	88.5 ± 1.5
.29	90.0	
.55	91.0	
.55	87.4	
.55	87.4	89.1 ± 1.7
.55	90.4	

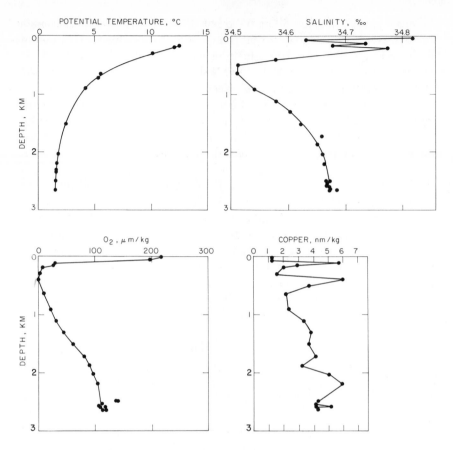

Figure 3. Profiles of potential temperature, salinity, dissolved oxygen, and copper for samples at 21°N, 109°W over East Pacific Rise

open ocean water without the use of larger sample volumes or a more sensitive instrumental method.

Tracer Experiments—Fe, Zn, Cd, Ni, Cu, and Mn. The success of the method for copper and lead suggested an extension to other metals. Preliminary experiments with iron, zinc, and cadmium showed poor reproducibility. In order to establish conditions suitable for the determination of these elements, experiments using radioactive tracers were done in the laboratory of M. P. Bender at the University of Rhode Island. The method was adapted to these experiments as follows:

1. Instead of glass fiber filters, 15 ml gooch crucibles with fine grade scintered discs were used

2. To compensate for the poorer retentivity of the sintered glass discs, the amounts of cobalt and APDC added were increased by a factor of eight. The samples were counted by placing the crucible directly over

a sodium iodide or germanium (lithium) detector. The standards were either tracer evaporated on a blotter disc placed inside a gooch crucible or tracer evaporated onto the sintered glass disc of a gooch crucible.

We estimate a 5% error in the calculated absolute recoveries. In the following experiments, the pH 1.8 solutions were unbuffered, the pH 4 and 6 solutions used an acetate buffer, and the pH 8 and higher solutions used an ammonia buffer.

DILUTE SOLUTIONS: ^{107}Cd, ^{64}Cu, ^{69}Zn, AND ^{65}Ni. These short-lived tracers were prepared by neutron activation of stable isotopes and added in solution to 100 ml of buffered distilled water. The pre-concentration and counting are described above. The results are shown in Figure 4. Nickel and copper are recovered in high yield throughout the range pH 1.8–8. Cadmium is recovered in high yield only at pH 1.8, and recovery decreases with increasing pH. Zinc is not quantitatively recovered at any pH although the yield increases with increasing pH.

DILUTE SOLUTIONS: ^{59}Fe^{III}, ^{65}Zn, AND ^{109}Cd. These long-lived tracers were added to 100 ml of buffered distilled water; the pre-concentration and counting are as described above. Figure 5 illustrates the results. As before cadmium is recovered in high yield at pH 1.8; the yield is less at higher pH except for the pH 6 sample. Zinc again is not quantitatively recovered at any pH. The yield is low below pH 4 and ∼ 80% above pH 4. Iron is recovered (> 85%) above pH 3.

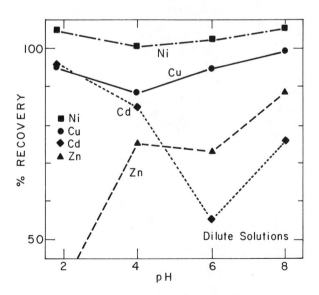

Figure 4.   Recovery efficiency as a function of pH for
^{107}*Cd,* ^{64}*Cu,* ^{69}*Zn, and* ^{65}*Ni in dilute solution*

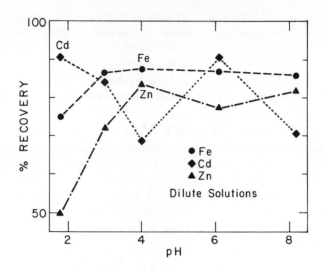

Figure 5.   Recovery efficiency as a function of pH for
⁵⁹Fe, ⁶⁵Zn, and ¹⁰⁹Cd in dilute solution

SEAWATER: ⁵⁹Feᴵᴵᴵ, ⁶⁵Zɴ, ᴀɴᴅ ¹⁰⁹Cᴅ. To check for possible changes in recovery arising from the ionic background in seawater, the above experiment was repeated for buffered seawater solutions. The results (Figure 6) were quite similar to those observed for distilled water.

MᴀɴɢᴀɴᴇSᴇ. Experiments similar to those above were done with ⁵⁴Mn, and the results are shown in Figure 7. Manganese recovery is very

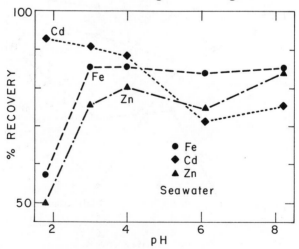

Figure 6.   Recovery efficiency as a function of pH for
⁵⁹Fe, ⁶⁵Zn, and ¹⁰⁹Cd in seawater

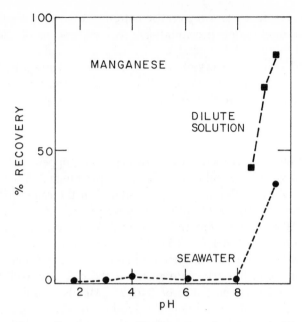

Figure 7. Recovery efficiency as a function of pH for ^{54}Mn in dilute solution and seawater

low in the pH range 1.8–6 and increases considerably above pH 8. The yield is high ($\sim 85>$) in dilute solution at pH 9.5 but lower ($\sim 35\%$) in seawater at the same pH.

REDISSOLUTION OF THE PRECIPITATE. In most of the above experiments the precipitate was dissolved from the disc by washing MEK–HNO₃ through the frit, and the cleaned crucible was recounted. The activity was reduced to background levels for nickel, copper, cadmium, and zinc, so dissolution by this method is effective. For iron, however, a substantial fraction of the original activity remained and was not removed by a second MEK-HNO₃ washing. Table II shows the results of such an experiment using ^{59}Fe as a tracer. The iron is incompletely removed by washing with the MEK–HNO₃ solution but is removed by a 3N HNO₃

Table II. Dissolving Iron Precipitate from Filters

^{59}Fe Original Activity (counts/4 min)	% Remaining after First MEK Wash	% Remaining after Second MEK Wash	% Remaining after First HNO₃ Wash	% Remaining after Second HNO₃ Wash
7124	28.8	12.4	0.98	0.36
13869	49.9	50.5	3.9	0.96
21528	43.5	43.9	4.7	1.3

wash. This result explains the erratic nature of early iron analysis by this method; despite a high precipitation recovery, not all of the iron was redissolved by the solution analyzed by atomic absorption.

ZINC. The low recovery of zinc is probably not caused by insufficient APDC since the concentration of APDC was increased by a factor of five, but with no increase in zinc recovery.

Discussion

The tracer experiments demonstrate that copper and nickel are precipitated with high yield throughout the pH range 2–8 using the conditions described earlier. Iron is precipitated in the range pH 3–8; cadmium is precipitated consistently only near pH 1.8. Zinc is precipitated in the 70–80% range from pH 3–8, and manganese is partially precipitated only at high pH (> 9.5). The falloff in iron and zinc recovery at low pH may arise from competition between the metal ions and hydrogen ions for the dithiocarbamate anion, as the pK_a of pyrrolidine dithiocarbamate has been reported to be 3.0 (9). Metals with a lower pyrrolidine dithiocarbamate conditional solubility product would be more severely affected by this mechanism; the reported stability sequence is $Zn < Cu < Ni < Pb < Cd$ (10). The generally lower zinc recovery overall may also be explained by this stability sequence. We have no definitive explanation for the low and erratic cadmium recovery at higher pH. Since the purpose of this paper is to present conditions for high yield co-precipitation rather than to explore the mechanism of co-precipitation, this aspect is worthy of further investigation.

Briefly, then, these experiments suggest the use of two slightly different precipitation procedures. One is useful for copper, nickel, and cadmium and is similar to that outlined in the methods section with the exclusion of buffer addition in order to maintain the sample near pH 1.8. The other method is useful for copper, nickel, lead, and iron and is similar to that outlined in the methods section except that a filter rinse with $3N$ nitric acid follows dissolution with the organic solvent. The combined organic and nitric acid washes are then evaporated to dryness over low heat, thus decomposing the chelate, and the residue is redissolved in $0.1N$ nitric acid prior to analysis.

Although the co-precipitation method was initially conceived with large volume extraction coupled with flame atomic absorption analysis in mind, its utility may prove to lie in other areas. In particular, the introduction of improved flameless atomizers will allow the determination of trace metals using much smaller sample volumes. We have combined pre-concentration by this method on 100 ml samples with use of a Perkin-Elmer HGA-2100 graphite furnace for the analysis of copper, nickel, and cadmium in seawater. For copper the sensitivity is comparable with that reported earlier for 4.1 flame analysis, and the precision is better. Alternatively, if lead rather than cobalt is used as the co-precipitating metal, neutron activation of precipitate can yield simultaneous analyses of several metals including cobalt (*11*). Radioisotopes can be concentrated by this method prior to counting; M. Bacon of the Woods Hole Oceanographic Institution is using this technique routinely for ^{210}Pb and ^{210}Po determinations.

Acknowledgments

We thank Peter Brewer and Derek Spencer for valuable advice during the development of the method, Michael Bender for suggesting the tracer experiments and making his laboratory facilities available for them, and Charles Lord for his work on the tracer experiments. We thank the scientific party, officers, and crew of RV Melville for their assistance.

Literature Cited

1. Brooks, R. R., Presley, B. J., Kaplan, I. R., "Determination of Copper in Saline Waters by Atomic Absorption Spectrophotometry combined with APDC–MIBK Extraction," *Anal. Chim. Acta* (1967) **38**, 321.
2. Brooks, R. R., Presley, B. J., Kaplan, I. R., "APDC–MIBK Extraction System for the Determination of Trace Elements in Saline Waters by Atomic Absorption Spectrophotometry," *Talanta* (1967) **14**, 809.
3. Brewer, P. G., Spencer, D. W., Smith, C. L., "Determination of Trace Metals in Seawater by Atomic Absorption Spectrophotometry," *At. Absorpt. Spectros.*, p. 70, ASTM STP 443, American Society for Testing and Materials, 1969.
4. Riley, J. P., "Analytical Chemistry of Seawater," "Chemical Oceanography," J. P. Riley, Ed., vol. **II**, p. 383, Academic, London, 1965.
5. Vogel, A. I., "A Textbook of Quantitative Inorganic Analysis," p. 485, John Wiley & Sons, New York, 1961.
6. Slavin, W., "Atomic Absorption Spectroscopy," p. 75, Interscience, 1968.
7. Kraus, K. A., Moore, G. E., "Anion Exchange Studies VI. The Divalent Transition Elements Manganese to Zinc in Hydrochloric Acid," *J. Am. Chem. Soc.* (1953) **75**, 1460.
8. Boyle, E. A., Edmond, J. M., "Copper in Surface Waters from Geosecs Stations in the Circumpolar Current South of New Zealand," *Nature* (1975) **253**, 107.
9. Hulanicki, A., "Complexation Reactions of Dithiocarbamates," *Talanta* (1967) **14**, 1371.
10. Arnac, M., Verboan, G., "Solubility Product Constants of Some Divalent Metal Ions with Ammonium Pyrrolidine Dithiocarbamate," *Anal. Chem.* (1974) **46**, 2059.
11. Brewer, P., Scranton, M., personal communication.

RECEIVED January 3, 1975. This work was supported in part by an NSF Fellowship. The laboratory and sea-going work were supplied by the Office of Naval Research. This is contribution No. 3 of the Geochemistry Collective at MIT.

7

Direct Determination of Trace Metals in Seawater by Flameless Atomic Absorption Spectrophotometry

DOUGLAS A. SEGAR and ADRIANA Y. CANTILLO

National Oceanic and Atmospheric Administration, Atlantic Oceanographic and Meteorological Laboratories, 15 Rickenbacker Causeway, Miami, Fla. 33149

Flameless atom reservoir atomic absorption spectrophotometry, because of its extremely high sensitivity, has found many applications in trace metal analysis of seawater, marine organisms, and sediments. Direct analysis of seawater for trace metals was not possible with early atomizer designs because of matrix interferences. A new generation of atomizer reduces these interferences and has been tested for its utility in direct analysis of seawater. All elements so far investigated—iron, manganese, copper, and cadmium —can be rapidly, simply, and precisely determined in their normal range of concentrations in seawater. Several precautions are necessary to obtain accurate results, as matrix composition, injection volume, atomizer conditions, and changes in graphite atomizer tube characteristics all affect the sensitivity of analysis.

The marine chemistry of trace transition metals is not well understood, despite many years of intense interest and research activity. The comparative lack of success of most investigations of marine geochemical cycles of transition elements undoubtedly arises largely from the inadequate analytical techniques used to determine elemental concentrations, particularly concentrations of metals dissolved in seawater. Not only are the historically preferred techniques inaccurate (1), but also their length and difficulty normally preclude the analysis of sufficient samples to describe adequately environmental variations. Unless these variations, both in time and space, can be adequately described, little can be learned about marine geochemical processes.

Trace metal concentrations in seawater are so low that contamination of the sample and loss of metal to container walls are critical problems in any analytical technique. These problems are particularly severe when the water sample must undergo extensive chemical treatment prior to the determination step. Most available techniques require such a chemical step or steps, because of their inadequate sensitivity and/or inability to determine the metal in the presence of the other sea water salts. Even those neutron activation procedures established for analysis of elements in seawater usually require a preactivation concentration step (2).

Recently, flameless atomization techniques have been developed for atomic spectroscopy, particularly for atomic absorption. Absolute sensitivities of atomic absorption using these flameless atomizers are, for most elements, comparable with or better than those attainable by any other technique. Additionally, unlike most other techniques the atomic absorption method is relatively free from interferences by other components of the sample matrix. Therefore, flameless atomic absorption holds great promise for direct analysis of trace metals in seawater and other environmental samples. This paper reports the successful application of a new design of commercial atomizer to direct analysis of several metals in seawater.

Flameless Atomizers

Flameless atomic absorption spectrophotometry is essentially very simple. A substrate upon which the sample matrix can be deposited is placed in or immediately adjacent to the spectrophotometer light beam, and a means of heating this substrate rapidly to 800°–3500°C is provided. Electrical resistance is usually the heating method used. The substrate

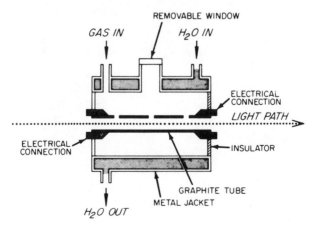

Figure 1.   Cross section of HGA-2000 atomizer head

itself has been made from various materials, and a large number of atomizer designs have been used. Two substrates—graphite and tantalum —appear to be best suited to routine use, and two basic atomizer designs —the open rod or West type and the closed furnace or Massman type— have been used (3). Of the two basic designs, the closed furnace appears to be preferable for routine analysis of most samples (3) because it does not require as stringent optical alignment as open filament types, generally can accept larger sample volume, and shows fewer inter-element interferences because of the smaller temperature gradient observed within the atomization zone. One such atomizer, the Perkin Elmer HGA-70 (later designated the HGA-2000) with a modified power supply, has been extensively evaluated for use in marine chemical analysis (4– 11). One of the major disadvantages of this atomizer was the physical arrangement which allowed the cooling, condensing atom gas cloud to remain in the optical path of the spectrophotometer while it was swept laterally out of the atomizer (Figure 1). This led to nonspecific absorption or scattering attenuation of the light beam, thereby preventing the analysis of high solid content matrices such as seawater (4). This attenuation was so great that it prevented the use of a background correction system, such as that based on the deuterium arc lamp (12). Recently, a new heated graphite atomizer has been designed, the Perkin Elmer HGA-2100 (Figure 2), which alleviates this problem by modifying the

Figure 2.   HGA-2100 atomizer head

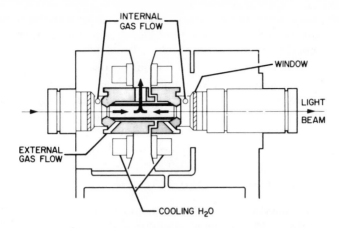

*Figure 3.   Cross section of HGA-2100 atomizer head
showing gas flow*

gas flow within the heated graphite tube to remove the hot atom cloud
from the light beam before the atom cloud is significantly cooled. The
purge gas enters the atomizer tube at each end and exits through the
sample introduction port at its center (Figure 3). A second inert gas
supply is provided outside the atomizer tube to prevent its oxidation.
The atomizer can accept up to 50 μl. of solution and be heated in three
different temperature steps up to about 3000°C.

Equipment

A Perkin Elmer model 503 atomic absorption spectrophotometer,
equipped with Perkin Elmer HGA-2100 heated graphite atomizer (Figure
2), a deuterium arc background corrector (*12*), and a strip chart recorder,
was used. The HGA-2100 graphite furnace was purged with argon.
Hollow cathode lamps were used except for cadmium for which an
electrodeless discharge lamp (Perkin Elmer) was used.

The reported temperature settings for the graphite furnace were
read from the HGA-2100 power supply readout and are approximate.
These temperatures are based upon the applied voltage across the
atomizer terminals. All absorbances were obtained from peak heights
read from either the strip chart or the digital peak height reader of the
Perkin Elmer 503.

The measurement of peak areas rather than peak heights would
undoubtedly eliminate or reduce some of the matrix effects on sensitivity
reported below (*13*). However, the integration mode of the Perkin Elmer
503 does not provide true peak area integration, but instead provides
signal averages for a preset time subsequent to initiation of the atomiza-

tion step. Because of the adverse signal-to-noise relationship caused by this procedure, the detection limits obtained with the integration mode are not as good as those obtained by peak height measurement. A fast response integrator programmed to the output peaks would undoubtedly enhance the analysis of complex samples such as seawater.

All standards were prepared by dilution of Alfa Inorganics Ventron primary standard solutions using acidified filtered surface Gulf Stream seawater (salinity ca. 36‰) or distilled water. Sample injections were made with Eppendorf microliter pipets with disposable plastic tips.

Preliminary Assessment of Seawater Analysis

The preliminary assessment of the behavior of seawater in the HGA-2100 was carried out in conjunction with the Perkin Elmer Co. Some of these results have been published elsewhere (*14*). The HGA-2100 gave rise to considerably smaller background absorbances than the HGA-2000 during atomization of sodium chloride solutions and measurement of the absorbance at the copper wavelength (324.7 nm). The charring temperature used was low enough so that no salt was volatilized before atomization. Plots of the molecular absorbance of sodium chloride vapor produced by the two atomizers as a function of concentration are shown in Figure 4. A 10 μl. aliquot of 35‰ seawater will contain 350 μg of total salt. From Figure 4 it can be seen that for a sample containing this quantity of sodium chloride, the background signal with the HGA-2000 is more than one absorbance unit while with the HGA-2100 it is about 0.1 absorbance —a value more readily correctable by means such as the deuterium arc background corrector (*12*). Even the reduced background absorbance afforded by the HGA-2100 is larger than desirable, particularly when analyzing samples having metal concentrations close to the detection limits of flameless atomic absorption. This condition is encountered for many elements in unpolluted seawater (Table I), and it is, therefore, necessary to use the selective volatilization technique where possible (*15*) to further reduce background interference.

Trace elements in seawater can be divided into two somewhat arbitrary groups according to their relative volatilities. The first group, including elements such as V, Co, Ni, Cu, Mn, Fe, Cr, and Mo is not volatilized at temperatures sufficient to volatilize the alkali chlorides. The second group consists of elements whose salts have volatilities similar to or greater than the alkali chlorides, including cadmium, zinc, lead, and gold. Selective volatilization can be used to remove the bulk of seawater salts prior to atomization of the low volatility elements but not the volatile elements (*4*). Elements which have been determined by flameless atomic absorption using the heated graphite atomizer are listed in Table I as being volatile or involatile. The division between the two groups is not well defined by observation, and some elements may fall into the other group when atomization from a seawater matrix is attempted as opposed to atomization of the simple salts. Table I shows approximate detection limits obtainable for various elements in simple aqueous solution. In addition, the approximate concentrations of the elements in unpolluted seawater are listed. A comparison reveals that, if detection limits comparable with those in distilled water can be obtained in seawater and if matrix effects can be compensated or eliminated, a number of elements could be determined by direct injection of seawater into the HGA-2100.

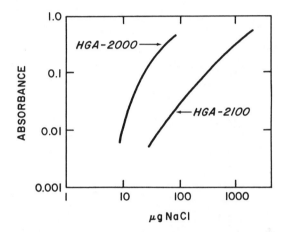

Figure 4. Absorbance of sodium chloride at 324.7 nm without background correction using the HGA-2000 and the HGA-2100 atomizers

An evaluation of direct analysis of seawater by the HGA-2100 was carried out for three elements with lower volatilities than the alkali metal chlorides (copper, iron, and manganese) and one element with higher volatility (cadmium). Analysis of seawater for each of these elements proved to be possible and sufficiently sensitive. However, a number of variables affect the analysis. These variables, in addition to the atomic absorption spectrophotometer settings, include the purge gas flow rate through the atomizer, the ashing temperature and time, the atomization temperature, the salinity of the sample, the volume of injection, and the changing surface properties of the graphite tube. To optimize the analytical sensitivity and precision, the effect of each of these variables was investigated.

Purge Gas Flow Rate. The purge gas flow rate can be adjusted up to about 220 ml of argon per minute. Normally the flow rate of the argon

Table I.   Trace Elements in Seawater and Detection Limits

Volatile Elements

Element	Approximate Detection Limit[a]	Approximate Seawater Concentration[b]
Ag	0.1	0.1
As	1	2.3
Au	0.5	0.005[c]
Bi	0.2	0.02
Cd	0.04	0.05
Hg	220	0.05[c]
In	16	0.0001
Pb	1	0.03[c]
Sb	5	0.01
Se	60	0.45
Sn	60	0.01
Te	600	—
Tl	3	0.01
Zn	0.02	5

[a] Detection limit in μg/l. for a 50-μl. injection. Detection limit taken to be equal to sensitivity listed by Perkin Elmer Corp. (21).
[b] From Riley and Chester (17) μg/l. for salinity = 35‰.
[c] Considerable variations known to occur.

gas is maintained as low as possible (about 50 ml/min) to maximize the residence time of atoms in the atomizer and, therefore, the peak atoms population and the analytical sensitivity. To obtain maximum sensitivities, the internal gas flow may even be switched off for a few seconds during the atomization step (16). Higher flow rates lead to generally lower sensitivities and are, therefore, undesirable. However, low flow rates will retard flushing of the cooling atom cloud from the furnace. When determining elements in a high salt content matrix, this significantly increases background absorption. Consequently, either the compensation ability of the deuterium arc background corrector is exceeded, or the reproducibility and precision of the analysis are reduced because of noise introduced by imperfect correction of large background signals. Thus, it was found that a flow rate of about 150 ml/min was optimal for manganese and copper analysis. Despite removal of the major seawater salts by ashing before atomization, sufficient matrix material remains to pro-

of Flameless Atomic Absorption Spectrophotometry

Involatile Elements

Element	Approximate Detection Limit[a]	Approximate Seawater Concentration[b]
Al	3	5[c]
Ba	6	30[c]
Be	0.7	0.0006
Co	2	0.08[c]
Cr	0.5	0.6[c]
Cs	2	0.5
Cu	1	3[c]
Dy	15	0.0009
Er	35	0.0009
Eu	800	0.0001
Fe	0.5	3[c]
Ga	50	0.03
Ir	60	—
Li	1	180
Mn	0.2	2[c]
Mo	2	10
Ni	3	2
Pd	3	—
Pt	2	—
Rb	1	120
Rh	4	0.01
Si	3	1000[c]
Sr	4	8500
Ti	40	1
V	7	1.5

duce significant background signals during atomization at low flow rates. Although sensitivity for manganese and copper is somewhat less at the chosen flow rate than at lower values, the difference is small and compensated for by improved reproducibility. For iron analysis, where a higher ashing temperature may be used and, therefore, more matrix material removed before atomization, the optimum flow rate is about 100 ml/min.

When atomizing cadmium from seawater, the atomic absorption signal is followed by a spurious non-atomic signal from the major salts (*see* Figure 22). The analysis depends upon the temporal separation of these two signals. At high gas flow rates, cadmium is swept out of the light beam before the spurious signal is generated. At lower rates, the sensitivity of the analysis is improved as the residence time of cadmium atoms in the light beam is increased. However, this increased residence time reduces the separation between the atomic and spurious peaks,

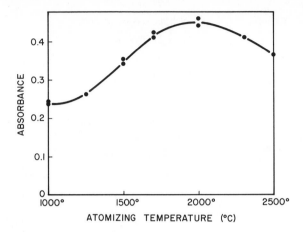

*Figure 5. Effect of atomization temperature on
the absorbance of 5 μl. of a 10-ppb spike of cadmium
in seawater*

causing overlap and interference. The optimum gas flow rate is, therefore, about 70 ml/min. It is possible that the limiting factor in resolution
of the cadmium and matrix signals is the relatively slow response time of
conventional atomic absorption spectrophotometer readout electronics.
If this is the case, then use of a faster readout system and lower gas flow
should improve sensitivity.

Atomization Temperature and Time. The maximum temperature of
the heated graphite tube during the atomization step and the rate at
which this temperature is achieved determines the rate of volatilization
and atomization of the sample and, therefore, the peak atom population
and sensitivity. For involatile elements, the peak height sensitivity increases with increasing temperature until a plateau is reached. The
optimum atomization temperature is then the lowest temperature at
which maximum sensitivity is obtained. For some volatile elements, the
peak absorbance may reach a maximum with increasing temperature and

Table II.   Optimum Conditions for

Element	Ashing Temp. (°C)	Ashing Time (sec)	Atomization Temp. (°C)
Cd	400	10	1500
Cu	600	25	2500
Fe	1250	25	2500
Mn	1100	25	2400

a Rotameter reading HGA-2100.

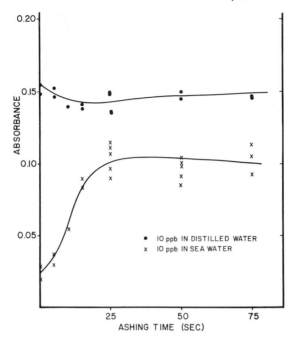

Figure 6.   Effect of ashing time on the absorbance of 40 μl. of a 10-ppb spike of manganese in distilled water and seawater (ashing temperature = 600°C)

may then decrease with further temperature increase (Figure 5). The optimum atomization temperatures for seawater analysis were found to be essentially the same as those for dilute aqueous metal salts and are listed in Table II.

The atomization time is set at the shortest time necessary for complete removal of the analysis element from the atomizer. Generally, a time is selected which continues atomization for a period after the peak signal is observed, corresponding to about twice the peak width at half height at the highest concentration to be determined. This ensures that

Seawater Analysis by Direct Injection

Atomization Time (sec)	Gas Flow[a]	Usual Sample Vol. (μl.)	Approximate Detection Limits (μg/kg)
7	40	10	0.01
7	80	50	0.5
7	60	20	0.4
7	80	20	0.3

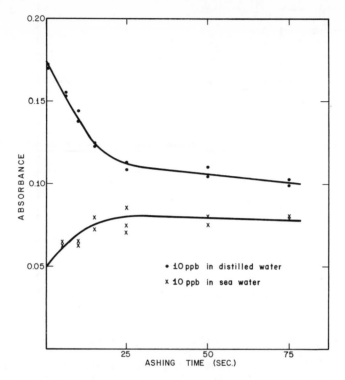

Figure 7. Effect of ashing time on the absorbance of 40 μl. of a 10-ppb spike of manganese in distilled water and seawater (ashing temperature = 1100°C)

memory effects are eliminated and permits examination of the analytical baseline immediately after the absorption signal while the atomizer is still at the atomization temperature. For some elements, particularly those whose analytical lines are of longer wavelength, there is a small but significant baseline shift caused by black body emission from the incandescent atomizer tube. This shift may be minimized by careful alignment of the optical train but still must be corrected for when low concentrations are determined.

Ashing Temperature and Time. The intermediate temperature heating cycle of the heated graphite atomizer, referred to here as the ashing cycle, removes as much of the matrix as possible without significant loss of the analyte. For involatile elements a significant proportion of seawater salts can be removed by this means before the atomization step. The choice of ashing temperature and time is made to obtain the optimum balance of sensitivity and reproducibility. At too low an ashing temperature or too short an ashing time the incomplete removal of the matrix salts and the consequent inability of the deuterium arc background corrector to compensate precisely for nonspecific absorption will reduce the reproducibility. Too high an ashing temperature will lead to significant loss of analyte metal from the atomizer before atomization and consequently a loss of analytical sensitivity. The effect of ashing time on manganese analysis in seawater is illustrated in Figures 6 and 7. At an ashing temperature of 600°C, little loss of manganese from the atomizer occurs even for long ashing times. However, the reproducibility of the analysis for seawater is extremely poor even at long ashing times (Figure 6). At an ashing temperature of 1100°C, although significant loss of manganese occurs from a distilled water matrix, the reproducibility of the analysis in seawater is much improved while the sensitivity is reduced by only about 25% (Figure 7). Optimum ashing times required at each temperature are similar. Little change in either reproducibility or sensitivity occurs with increasing time above 25 sec.

The effect of ashing temperature upon the analysis of iron, manganese, and copper is illustrated in Figures 8, 9, and 10, respectively.

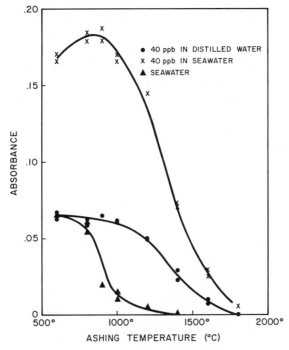

Figure 8. Effect of ashing temperature on the absorbance of 20-μl. injections of a 40-ppb spike of iron in distilled water and seawater and of unspiked seawater (Fe < 0.5 ppb)

When introduced in chloride salt solution in distilled water, the response to changes in ashing temperature is, in each instance, relatively simple. The analytical sensitivity falls off at temperatures above 500°C, as increasing amounts of metal are lost from the atomizer during the ashing cycle. When the salts are introduced in natural seawater, observed sensitivity changes are more complex. As with distilled water, sensitivity drops above a critical temperature, which is different for each element, because of loss of the element from the atomizer during ashing. However, below this temperature, the sensitivity drops instead of leveling off as with simple solutions. The cause of this sensitivity loss is unknown, although it must be caused by either lowered instrument response arising from large nonspecific absorption and considerably decreased light level reaching the photomultiplier or, more likely, chemical interference by the major seawater salts. Such chemical interference might be caused by suppression of dissociation of molecular species of the analyte element in the molecule and atom cloud by the presence of large quantities of more easily dissociable salts. This would be analogous to the suppression of ionization, achieved for many elements in flames or arcs by the addition of large quantities of easily ionizable elements. Sodium chloride at a concentration of 3.5 g/l. has a larger suppression effect than seawater with a total salt content of 3.5 g/l. The effect is thus not simply determined by the total quantity of elements in the sample but is also dependent upon the composition of the matrix. The complexity of the

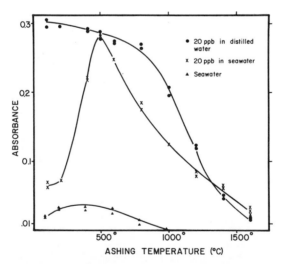

Figure 9. Effect of ashing temperature on the absorbance of 40-μl. injections of 20-ppb spike of manganese in distilled water and seawater and of unspiked seawater (Mn < 1 ppb)

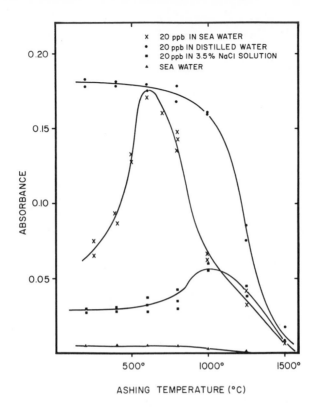

Figure 10.   *Effect of ashing temperature on the absorbance of 40-μl. injections of a 20-ppb spike of copper in distilled water, seawater, 3.5% sodium chloride solution, and of unspiked seawater (Cu < 0.5 ppb)*

atomization phenomenon from a complicated matrix is further illustrated by the observation that, although the managanese and copper sensitivities are suppressed in seawater as compared with simple chloride solutions regardless of ashing temperature, the sensitivity for iron is considerably enhanced in seawater. Too little is known about the chemistry of atomic vapor clouds, such as are generated in the heated graphite atomizer, to enable more than speculation upon the cause of enhancement or suppression. However, the matrix clearly must affect such vital parameters as the chemical form of the analysis element deposited in the solid state after drying in the atomizer and the volatilization and dissociation of these compounds. Considerably more research, both experimental and theoretical, is called for in this area.

Injection Volume. The volume of sample injected into the HGA-2100 may be up to 100 μl., but usually is between 10 and 50 μl. The

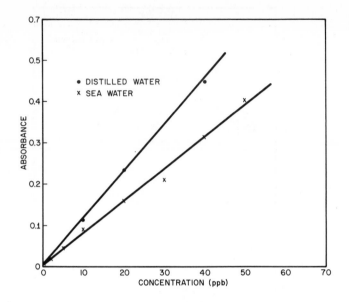

*Figure 11.   Calibration for manganese (0–50 ppb) in seawater
and distilled water (injection volume = 40 μl.)*

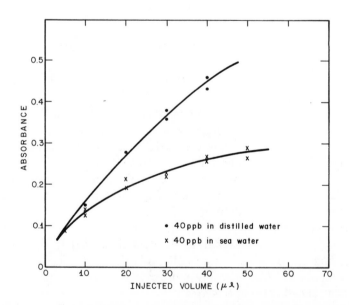

*Figure 12.   Calibration for manganese (40 ppb) in seawater
and distilled water (injection volume = 10–50 μl.)*

injection volume affects analytical sensitivity both with simple salt solutions and with seawater. Figures 11 and 12 both show calibration curves for manganese in distilled and seawater. Figure 11 shows linear calibrations obtained by injecting different concentrations of manganese in identical volumes of sample. Figure 12 was obtained by injection of different volumes of a single concentration of manganese in both distilled and seawater. As the injection volume increases, the peak height drops, in each instance leading to curvature of the calibration. With distilled water injections, this curvature is probably caused by a change in the volatilization rate of manganese because of its wider distribution on the

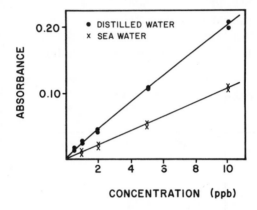

Figure 13.   Calibration for cadmium (0–10 ppb) in seawater and distilled water (injection volume = 5 μl.)

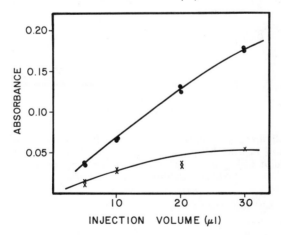

Figure 14.   Calibration for cadmium (0.5 ppb) in seawater and distilled water (injection volume = 5–30 μl.)

floor of the graphite tube, which is not uniformly heated. Atomization does not take place simultaneously at all points in the tube, which leads to a broader, smaller, output signal. In seawater, the curvature is much greater, and the large quantity of salt must have an additional effect at larger injection volumes. Similar calibrations were obtained for other elements. Figures 13 and 14 show the corresponding calibration for the volatile element cadmium.

In order to show the effect of total salt quantity in the atomizer, a series of injections were made for cadmium and manganese analysis with different volumes of solution but with the same total quantity of the analysis metal present per injection. Three series of injections were made—in distilled water, in seawater, and in seawater diluted to maintain the total salt quantity per injection constant. The results are shown in Figures 15 and 16 for manganese and cadmium, respectively. It is

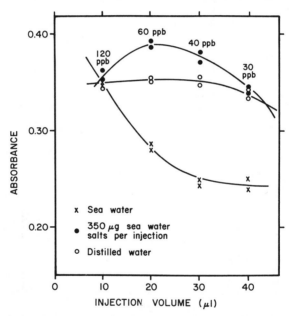

Figure 15. Effect of injection volume on the absorbance of 1.2 ng of manganese in distilled water with 350 ng of seawater dissolved salts per injection and in seawater of 35‰ salinity

apparent that the injection volume alone has only a small effect on the sensitivity, although some sensitivity loss occurs with increasing volume of distilled water. The effect of increased total salt content in the atomizer is to reduce the sensitivity in each case, presumably because of a suppression of dissociation or similar phenomenon. The effect of maintaining

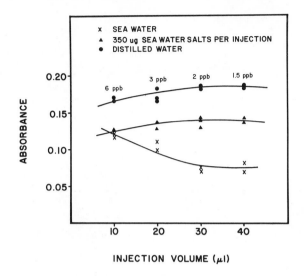

Figure 16. *Effect of injection volume on the absorbance of 60 ng of cadmium in distilled water with 350 ng of seawater dissolved salts per injection and in seawater of 35‰ salinity*

constant salt quantity and constant metal quantity while varying the injection volume is complex but resembles to some extent the effect of salinity (*see* Figures 17 and 20).

Although at this time the sensitivity variations seen when changing injection volume with saline samples cannot be explained, it is clear that

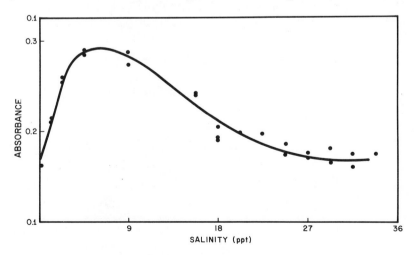

Figure 17. *Effect of salinity on the absorbance of 40 µl. of 20 ppb manganese in seawater*

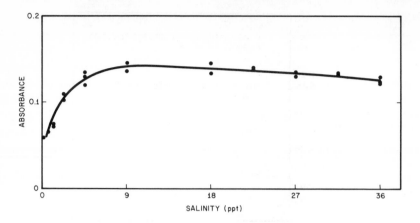

Figure 18. Effect of salinity on the absorbance of 20 μl. of 40 ppb of iron in seawater

for accurate results all samples and standards must be injected in the same volumes and with the same salinity.

Salinity. The sensitivity of the analysis for each of the elements investigated depends on the salinity of the sample (Figures 17–20). Sensitivities for iron and manganese are both enhanced at low salinities, compared with distilled water standards, and in each instance sensitivity falls off at higher salinities. The effect of salinity on copper and cadmium analysis is more complex (Figures 19 and 20). A large drop in sensitivity occurs from distilled water to low salinities. At higher salinities, the sensitivity increases again and then drops slowly. As has already been stated, it is not possible to explain variations of sensitivity with salinity

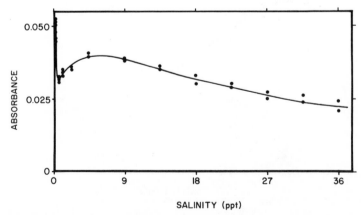

Figure 19.   Effect of salinity on the absorbance of 40 μl. of 20 ppb of copper in seawater

because of the lack of knowledge of the chemistry of hot atomic and molecular clouds.

The maximum rate of change in sensitivity for each element takes place at salinities near those of fresh waters; therefore, such samples should always be analyzed by standard additions. This is also necessary because of the variability of major ion compositions of natural fresh water, which may be expected to affect sensitivity, along with changes in the total salt content. Fortunately, small changes of salt content near the values of salinity found in the open sea have very little effect on the analytical sensitivity for any of the metals studied. Trace metal analysis of seawater may, therefore, be performed using standard additions on selected samples only.

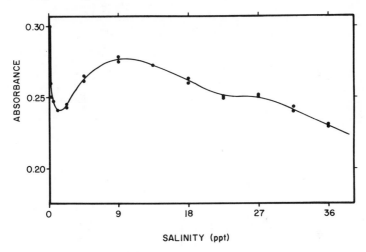

Figure 20.   Effect of salinity on the absorbance of 10 μl. of 20 ppb of cadmium in seawater

Seawater Analysis, Analytical Conditions, and Procedure

Analytical conditions adopted for analysis of iron, manganese, cadmium, and copper in seawater are summarized in Table II. Output peaks obtained are illustrated in Figure 21 for copper and Figure 22 for cadmium. For copper and other refractory elements, spurious signals generated by atomizing seawater salts at the ashing temperature are not recorded since the recorder is switched on automatically immediately prior to the atomization step. However, immediately following the atomic absorption peak for cadmium, the atomized major salts produce strong scattering and molecular absorption which reduces the light intensity passing through the atomizer almost to zero. This leads to spurious signals on the recorder which at first are negative, then positive, and

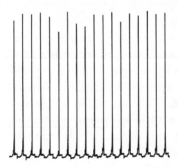

Figure 21. Reproducibility of analysis of copper in seawater, 50 μl. injections of a 40-ppb spiked sample of 35‰ salinity (recorder scale expansion, 2×; chart speed, 5 mm/min)

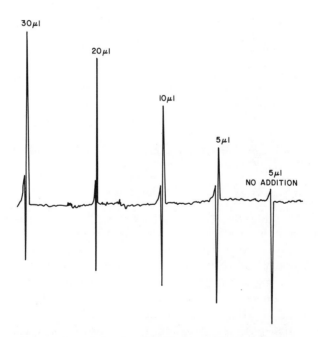

Figure 22. Recorder signal for cadmium analysis in seawater and seawater spiked with 0.5 ppb of cadmium (recorder scale expansion, 5×; chart speed, 160 mm/min)

then return to the baseline during the atomization cycle (Figure 22). These spurious peaks may be ignored and do not affect the analytical signal as long as the electrodeless discharge lamp and deuterium arc lamp beams are well aligned and intensity matched. Examination of the response at a nonabsorbing line close to the analytical line shows that the cadmium peak precedes, and is unaffected by, the scattering signal. Calibration curves for iron, copper, cadmium, and manganese are shown in Figures 23, 24, 25, and 11 respectively. From these calibrations, it can

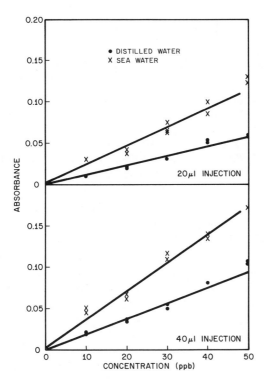

*Figure 23.   Calibration for iron (0–40 ppb)
in seawater and distilled water*

be seen that analysis of samples of seawaters with concentrations of these elements within the normal range (17) is possible with acceptable precision. Precision is estimated to be better than ±10% above about 0.1 ppb cadmium and above about 2 ppb for the other elements. The 18 successive injections of 40 ppb of copper in seawater shown in Figure 21 have a standard deviation of ±4%. The analytical procedure is, therefore, very simple. Analytical conditions are set, and an appropriate volume of acidified seawater in injected.

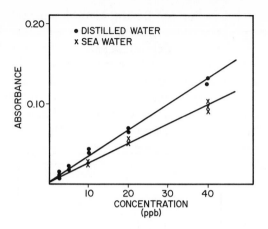

Figure 24.   Calibration for copper (0–40 ppb)
in seawater and distilled water (50-µl. injection)

A number of precautions must be observed to obtain accurate data. These are necessitated primarily by variability in graphite tubes and degradation of these tubes during use. No two tubes have precisely the same surface properties and, for example, variations of sensitivity with salinity (Figures 17 through 20) are completely reproducible only in form and not in absolute magnitude from tube to tube. In addition, sensitivity declines slowly with tube use as the tube surface is degraded. Fortunately the sensitivity loss is linear for at least the early part of a graphite tube life (Figures 26 and 27) and can be easily calibrated. Loss of sensitivity increases rapidly and reproducibility decreases dramatically at the end of the tubes useful life. Tubes are discarded before

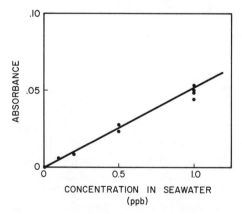

Figure 25.   Calibration for cadmium (0–1
ppb) in seawater (5-µl. injection)

this rapid deterioration begins. Tube lifetimes vary slightly from one to another, but vary more with changes in sample matrix, and analytical conditions. The following steps are, therefore, adopted to ensure accurate results.

1. Each sample is injected at least twice or until replication of output peaks is better than a desired precision level (usually $\pm 10\%$).

2. A series of standards of various concentrations, spiked into one of the seawater samples, is run to establish a calibration curve with each new tube.

3. An intermediate concentration standard is reanalyzed after approximately every 20 injections, to calibrate sensitivity drift.

4. Samples are grouped within salinity ranges covering no more than 5‰ (smaller for low salinity samples). Spiked standards are prepared in representative seawater samples from each of these ranges.

5. Sample and standard injection volumes are always the same.

Direct injection analysis for iron, manganese, copper, and cadmium has been successfully performed on more than 500 samples of seawater taken in the New York Bight as part of the National Oceanic and Atmospheric Administration's (NOAA) Marine Ecosystem Analysis (MESA) Program. Although the instrumental analysis is tedious, no problems have been incurred in this routine process. Approximately 50–100 samples may be analyzed for one element in a day.

Future Developments

The four elements investigated here are by no means the only elements which may be directly determined in seawater. Preliminary investigations for other elements, including As, Pb, Zn, Si, and Al, are encouraging and will be pursued further. In addition, for those elements that cannot be directly determined, pre-concentration may be carried out before analysis, although this is usually tedious and may lead to contamination error (6).

Biological tissues and sediments may be analyzed for many trace elements by flameless atomic absorption after dissolution (6, 7). The reduced background signal observed with the new HGA-2100, compared with its predecessors, suggests that direct analysis for many of these elements may now be possible in these matrices. The sample may be introduced either directly as solids (18) or, more easily, as homogeneous suspensions after ultrasonic dispersion in water or solubilizing media (19). Perhaps the most exciting future prospect for the flameless atomizer in chemical oceanography is its use as a high sensitivity specific detector for gas and eventually liquid chromatography (20). Such instruments may provide the means for determining metallo-organics such as the

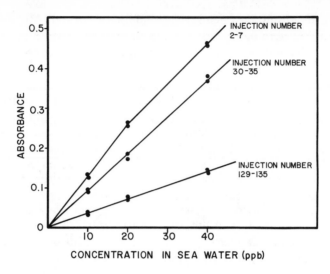

*Figure 26.   Change of calibration for manganese (10–40
ppb) in seawater with graphite tube use*

alkylated lead, mercury, and cadmium compounds in environmental
samples.

Analysis of seawater for zinc, chromium, and nickel has been carried
out successfully by the techniques outlined in this paper. Approximate
detection limits are 0.05 ppb for zinc, 0.5 ppb for chromium, and 3.0 ppb
for nickel.

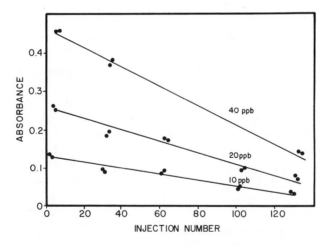

*Figure 27.   Linearity of calibration drift for manganese
(10–40 ppb) in seawater with graphite tube use*

The strength of compression used to hold the graphite tube in place in the HGA-2100 atomizer head critically affects the tube life. Optimization of this parameter has resulted in considerably slower sensitivity degradation than shown in Figures 26 and 27. The sensitivity degradation is nevertheless linear as shown.

Literature Cited

1. Brewer, P. G., Spencer, D. W., "Technical Report, Woods Hole Oceanographic Institute" (1970) **70–62.**
2. Goldberg, E. D., "Marine Pollution Monitoring: Strategies for a National Program," NOAA, Washington, D. C., 1972.
3. Kirkbright, G. F., *Analyst* (1971) **96** (1146), 609.
4. Segar, D. A., Gonzalez, J. G., *Anal. Chim. Acta* (1972) **58,** 7.
5. Segar, D. A., *Proc. 3rd Int. At. Absorp. Fluores. Spectrom. Cong., Oct., 1971,* Paris, France (1973) 523.
6. Segar, D. A., *Int. J. Environ. Anal. Chem.* (1973) **3,** 107.
7. Segar, D. A., Gilio, J. L., *Int. J. Environ. Anal. Chem.* (1973) **2,** 291.
8. Paus, P. E., *At. Absorpt. Newsl.* (1971) **10,** 69.
9. Paus, P. E., *Z. Anal. Chem.* (1973) **264,** 118.
10. Muzzarelli, R. A. A., Rocchetti, R., *Anal. Chim. Acta* (1973) **64,** 371.
11. Robinson, J. W., Wolcott, D. K., Slevin, P. J., Hindman, G. D., *Anal. Chim. Acta* (1973) **66,** 13.
12. Kahn, H. L., Manning, D. L., *Am. Lab.* (1972) **4,** 51.
13. L'vov, B. V., "Atomic Absorption Spectrochemical Analysis," p. 324, American Elsevier, New York, 1970.
14. Ediger, R. D., Peterson, G. E., Kerber, J. D., *At. Absorp. News.* (1974) **13,** 61.
15. Segar, D. A., Gonzalez, J. G., *At. Absorp. Newsl.* (1971) **10,** 94.
16. Kahn, H. L., Slavin, S., *At. Absorp. Newsl.* (1971) **10,** 125.
17. Riley, J. P., Chester, R., "Introduction to Marine Chemistry," Academic, London and New York, 1971.
18. Kerber, J. D., Koch, A., Peterson, G. E., *At. Absorp. Newsl.* (1973) **12,** 104.
19. Segar, D. A., *Abstracts 4th Int. Conf. At. Spectros.,* Toronto, Canada, 1973.
20. Segar, D. A., *Anal. Lett.* (1974) **7,** 89.
21. Perkin Elmer Corporation, "Analytical Methods for Atomic Absorption Spectroscopy Using the HGA Graphite Furnace," Perkin Elmer, Conn., 1973.

RECEIVED January 13, 1975. Support for this research has come primarily from National Oceanic and Atmospheric Administration's (NOAA) Marine Ecosystem Analysis (MESA) New York Bight Program with some support from NSF grant GA 33003. The Environmental Research Laboratories of the National Oceanic and Atmospheric Administration does not approve, recommend, or endorse any product; and the results reported in this document shall not be used in advertising or sales promotion or in any manner to indicate, either implicitly or explicitly, endorsement by the United States Government of any specific product or manufacturer.

8

Automated Anodic Stripping Voltammetry for the Measurement of Copper, Zinc, Cadmium, and Lead in Seawater

ALBERTO ZIRINO

Chemistry and Environmental Sciences Division, Naval Undersea Center, San Diego, Calif. 92132

STEPHEN H. LIEBERMAN

Chemistry and Environmental Sciences Division, Naval Undersea Center, San Diego, Calif. 92132 and Department of Chemistry, California State University, San Diego, San Diego, Calif. 92115

Copper, zinc, cadmium, and lead in seawater are analyzed by automated anodic stripping voltammetry (AASV) using a tubular mercury–graphite electrode. The system is controlled by a specially designed solid state programmer that controls the potentiostat, two peristaltic pumps, and five solenoid-operated Teflon valves and alternately circulates a mercury solution, a seawater sample, or a standard solution through the electrode assemblage. Automated analyses with standard additions are made either by the linear sweep or differential pulse process. The system has been evaluated extensively in the field, where the electrode performed satisfactorily for over 100 analyses without reconditioning. Seawater concentrations of trace metals measured with the automated system generally agree with those available in the literature.

Automated analysis systems are necessary for large-scale marine surveys. Anodic stripping voltammetry (ASV) with the tubular mercury graphite electrode (TMGE) possesses adequate sensitivity and precision under repeated use to characterize zinc in San Diego Bay water. The TMGE, made by electrolysis of a mercuric nitrate solution to form a thin mercury film inside a graphite tube, is described elsewhere (1).

In this work we have automated the TMGE–ASV system and have extended the procedure to determine copper, cadmium, and lead as well as zinc in marine waters.

The automated system is controlled by a specially designed programmer which controls the potentiostat, the recorder, two peristaltic pumps, and five solenoid-operated valves which alternately circulate a mercuric nitrate solution, a seawater sample, or a standard solution through the electrode assemblage. Automated analyses for copper, zinc, cadmium, and lead were made on the pier of the Scripps Institution of Oceanography, San Diego, Calif., and aboard ship in Puget Sound, Wash. This report describes the automated system, discusses the performance of the electrode under continued use, and presents the results obtained in the field.

Experimental

Instrumentation. Early work was performed with a specially built, cam-operated, electromechanical timer coupled to a modified Heathkit polarographic analyzer. Subsequently, a solid-state programmer system was designed and constructed. This system consists of a programmer unit, a relay unit, and connecting cables. The programmer unit contains the sequencing and control logic while the relay unit contains the out-power relays. Control and timing are implemented with TTL integrated circuits and compatible output devices. The programmer is coupled to a PAR 174 polarographic analyzer which allows determinations to be made by differential-pulse anodic stripping voltammetry (DPASV). Current–voltage displays are recorded on a modified Sargent-Welch model SRG strip chart recorder which is also under program control. This system is illustrated in block form in Figure 1.

The programmer unit is an upright cabinet containing an isolated power supply, timing and control logic circuits, operating controls, and program status displays. The sequence and duration of the program instructions are selectable with front panel switches. The full program sequence includes the following:

1. Introduction of the sample,
2. Application of the mercury film,
3. Electrolysis at constant potential of the mercury film and the sample,
4. Oxidation of the mercury film,
5. Replacement of the sample.

This sequence is a one-program cycle and may be an analysis of the sample or of the standard solution. A sample can be analyzed one to nine times, and a display shows the number of the cycle in progress.

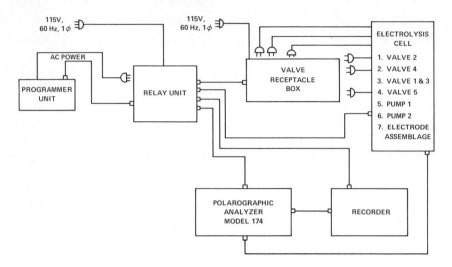

Figure 1.   Block diagram of automated ASV system

At the end of the selected number of sample analysis cycles, one analysis of the standard is performed. The system then either terminates work or repeats the entire sequence at the discretion of the operator. The interval between analyses is under program control and may vary from 2 min to 1 hr. A provision for making standard additions under program control has also been made, but as of this writing the automated pipette has not been installed, and standard additions were made with a hand-held micropipette. Details of the programmer system and wiring diagrams are available upon request.

Electrolysis Cell. The electrolysis cell consists of the TMGE, three solution reservoirs, and a peristaltic pump. A schematic of the cell is shown in Figure 2. The reservoirs which contain the mercuric nitrate and the standard solution are 1-l. FEP Teflon bottles, and the sample reservoir is a 250-ml FEP Teflon bottle. Each of these may be circulated through the electrodes by appropriate switching of four solenoid-operated Teflon valves (Valcor Engineering Corp.) which are numbered 1 through 4 in Figure 2. Valves and reservoirs are interconnected with Teflon tubing of 3/16-in. id and nylon fittings (Swagelock, Inc.). The pump is a variable-speed Masterflex peristaltic pump (Cole Parmer No. 7545-15); formula-3603 Tygon tubing of 3/16-in. id and 3/8-in. od is used in the pump head.

The cell has been designed to rinse automatically and to obtain a discrete sample of an appropriate volume from a continuously flowing stream. By means of an additional peristaltic pump identical to that used in the electrode assemblage, seawater is continuously pumped at a high rate (250 ml/min) through a three-way Teflon valve (No. 5 in Figure 1) to a drain. Activation of the valve causes the stream to be diverted into the sample reservoir through an opening at the bottom. The sample fills the vessel and drains into the drain line through an opening on the side of the reservoir. The volume of the sample is controlled by the position of this outlet. Samples of 100 ml and 200 ml were used in this work. The extent of rinsing of the reservoir is controlled by the length of time valve 5 is activated. Oxygen is continually removed from the sample and from the other solutions by sparging with prepurified nitrogen, or carbon dioxide, or a mixture of the two (2).

Continuous Flow and Batch Sampling. With appropriate modifications of the cell, a stream of seawater may be passed directly through the electrode assemblage and into the discharge line. This eliminates recycling the sample in the reservoir but requires efficient removal of oxygen from the stream. A system of this type was used to conduct a survey in the Gulf of Mexico where oxygen was removed from the seawater stream by a specially designed counter-current exchange apparatus. After this trial, however, direct continuous-flow voltammetry was discontinued because it was difficult to calibrate in terms of concentrations, and it was suspected that considerable adsorption of metals

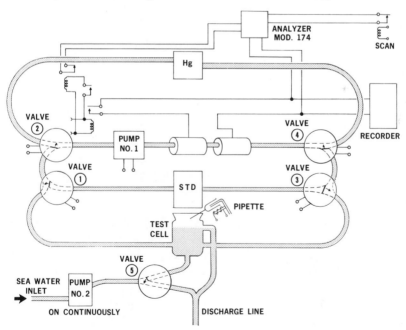

SEA WATER ANALYSIS RELAY AND VALVE CONTROL

Figure 2.   Schematic of electrolysis cell

occurred on the plates of the counter-current exchanger. Batch sampling from a flowing stream was thus adopted. Unfortunately, this mode of sampling also has inherent problems. Principally, it was observed that the hold-up volume of the sample in the electrode assemblage mixed with the succeeding sample. Solid products resulting from the oxidation of the mercury film, such as mercurous chloride (3), also tended to be removed from the electrode and deposited into the succeeding sample by flow-produced friction. However, the carry-over of metals from the previous analysis is a serious problem only when successive samples are of disparate concentrations (as after a standard addition) or when sample volumes are kept very small. Under the latter circumstances the carry-over error may be as much as 20%, and analyses following measurement of samples with high concentration must be disregarded.

Electrodes. The procedure for preparing the TMGE remains essentially as reported earlier (1). However, we have noted that it is no longer necessary to maintain the impregnated electrodes under vacuum prior to sanding and insertion into the system. Also, as of the final writing of this report, we observed that silver contamination from the reference electrode (1) seriously interfered with the copper determination. We have therefore tried to identify those results which would be affected by the presence of silver in the sample. Finally, we have replaced the contaminating electrode with a newly designed tubular silver/silver chloride-platinum reference-counter combination electrode. This and other improvements to the system will be presented in a subsequent communication.

Stock Solutions and Standard Additions. One-liter, $10^{-3}M$ stock solutions were made by dissolving the appropriate amounts of copper, zinc, cadmium, and lead or their salts in nitric acid and diluting to volume. From these, two sets of standard solutions were prepared by dilution with deionized, quartz-distilled water: a quadruple standard, used in the zinc analysis, containing $5.0 \times 10^{-6}M$ copper, $2.5 \times 10^{-5}M$ zinc, and $2.5 \times 10^{-6}M$ cadmium and lead; and a triple standard, $2.0 \times 10^{-5}M$ in copper, $1.0 \times 10^{-5}M$ in cadmium, and $2.0 \times 10^{-6}M$ in lead, used in the analysis of these three metals. Additions of 100 μl. or multiples thereof were made to 100- or 200-ml seawater samples.

Analytical Procedure. For most work, water was sampled on board while underway by means of the already-mentioned second peristaltic pump connected to heavy-walled, 1/4-in. id Tygon tubing. Approximately 250 ft of Tygon tubing were required to bring near-surface water samples to the ship's laboratory. Only about 50 ft of tubing were required at a stationary pier location. Absence of adsorption of trace metals or contamination by the Tygon tubing was verified aboard ship by simultaneous sampling over the side with a Teflon beaker tied to a cotton line. No contamination or adsorption of trace metals by the tubing was detected after several hours of use.

In general, a discrete analysis is performed, as follows: the sample reservoir is rinsed by activating valve 5 for 2–4 min, and a sample is introduced and sparged with nitrogen, or carbon dioxide, or a mixture of the two. The mercury solution is circulated through the electrode assemblage, and a potential is applied to the electrode to deposit a mercury film. The mercury deposition potential is also the deposition potential of the sample. We chose −1.4 V *vs.* silver/silver chloride for the electrolysis of zinc and −1.0 V *vs.* silver/silver chloride for the concurrent deposition of cadmium, lead, and copper. After deposition of the film the sample is introduced into the system by activating valves 2 and 4. In order to limit mixing of the mercury solution with the sample, valve 4 is activated a few seconds after valve 2; the actual duration of the delay depends on the flow rate in the cell. Constant-potential electrolysis of the sample is carried out for a predetermined time, usually 4–10 min. Then the flow is stopped, and the trace metals accumulated in the film, as well as the film itself, are oxidized by applying a linearly varying or pulsed potential ramp to the electrode. The anodic potential scan is terminated after the oxidation of the mercury film (at least +0.5 V *vs.* silver/silver chloride). The oxidation current is automatically recorded. Flow is then resumed, mercury solution is reintroduced into the electrode assemblage, and the sample is replaced in the cell. The entire procedure can then be repeated at the discretion of the operator. A flow rate of 160 ml/l. was used for all of the work described herein.

Results and Discussion

Electrode Performance. For the determination of copper, cadmium, and lead, the performance of the electrode was evaluated by continually recycling a 2-l. sample of San Diego Bay water for 129 consecutive analyses. During the last 22 determinations standard additions of copper, cadmium, and lead were made every third analysis. These additions increased the concentration of copper, cadmium, and lead in the sample by 1.0, 1.0, and 0.4 ppb respectively. All analyses were made by DPASV at pH 4.9 by sparging with carbon dioxide. The results of the consecutive determinations of copper and lead are presented in Figure 3. The relative heights of the trace metal peak currents are plotted on the ordinate as functions of the number of analytical determinations. The cadmium data are similar to that of copper and lead and have been omitted from the figure for clarity.

Copper, cadmium, and lead peak currents appear to increase for approximately 20 determinations and then decrease steadily thereafter. This early increase in peak currents is related to an increase in activity or coverage of the mercury film with successive determinations and is similar to the observations reported earlier (1) although it occurs under different experimental conditions. Following 100 consecutive analyses there is a 70% decrease in the copper peak current, a 60% diminution for lead current, and a 40% reduction in the cadmium current. An

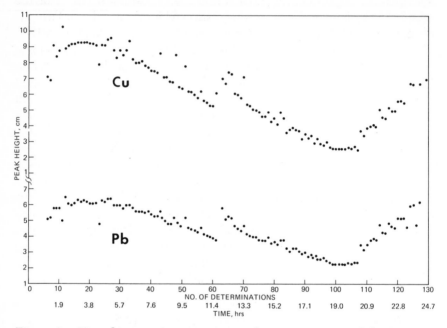

Figure 3.   Time history of consecutive analyses of copper and lead in San Diego Bay water

abrupt increase in sensitivity which extends the useful life of the electrode occurs at the 62nd cycle. This enhancement is related to the oxidation of the mercury film in a carbon dioxide bubble trapped in the TMGE. We do not understand why the quality of the film should be enhanced by this.

We have also studied the performance of the electrode by carrying out 380 consecutive analyses for zinc in San Diego Bay water. A fresh sample was used for each analysis, and although zinc was of primary interest, copper, cadmium, and lead current peaks could also be observed. The sensitivity of the electrode to zinc decreased slightly after the first 20 determinations and then remained relatively constant even after five days of continued use. Copper, cadmium, and lead peak currents increased initially, then decreased as previously described. After about 100 determinations they were no longer useful for quantitative analysis.

At present we do not fully understand why the electrode gradually loses sensitivity to copper, cadmium, and lead. Presumably the adsorption of organic and inorganic materials from seawater as well as the reduction and accumulation of interfering trace metals decrease sensitivity. Some of these problems are discussed in the next section.

Interferences. Practically, interferences on the TMGE can be thought of as consisting of two types: those which affect the electrode by causing peak broadening and loss of sensitivity but which do not interfere with a quantitative determination by standard addition and those which produce measurable peak currents at the same potential as metals undergoing determination. The latter cannot be compensated for by standard addition. The successful application of the AASV technique to oceanography depends on containing interferences of the second kind.

An initial concept of the interferences which may occur at the TMGE can be obtained from the work of Smith and Redmond (4), who studied the stripping characteristics of several trace metals at the hanging mercury drop electrode (HMDE) in a seawater medium. Because trace metals are much less concentrated in the HMDE than in the TMGE, Smith and Redmond's work probably suggests the minimum interferences that can be expected with the TMGE. They observed that nickel, antimony, and zinc produced current peaks at the "zinc" potential, cadmium and tin oxidized at "cadmium" potential, and copper, nickel, and vanadium oxidized at the "copper" potential while lead appeared to be free from interferences. Zirino and Healy (2), however, pointed out that tin could also interfere with the lead determination.

Fortunately, at equal concentrations, nickel peak currents were only one-seventh those of zinc and one-ninth those of copper. Antimony, besides interfering with zinc, also produced a second, clearly identifiable peak at −0.28 V *vs.* the standard calomel electrode. By observing that this peak could not be obtained with natural seawater samples and by considering that antimony peak currents were only one-sixth those of zinc, Smith and Redmond concluded that the concentration of antimony in seawater was too low to cause problems. Zirino and Healy (2) also concluded that tin hydrolyzed too readily in seawater to interfere in the determination of lead. Finally, the relative insensitivity of the HMDE to vanadium as well as its low concentration in seawater made interferences from this element unlikely.

A systematic study of interferences at the thin film electrode was made by Seitz (5), who observed that nickel and zinc interfered with the copper determination by causing the peak current to broaden. Also, the nickel–copper intermetallic compound did not strip out of the mercury film. These interferences could be easily avoided, however, if the electrolysis of the sample were carried out at potentials too cathodic to reduce nickel and zinc. A more serious problem was that silver oxidized with copper and was indistinguishable from it. Moreover, silver cannot be easily removed from the sample, and thus the extent of silver interference in the determination of copper depends on the relative quantities of these two metals in the sample. For seawater, however, the expected silver concentration is approximately one-tenth of the copper concentration (6, 7).

Copper, nickel, and cobalt were found by Seitz (5) to diminish the height of the zinc current peak by broadening it. Although the concentration of cobalt in seawater was deemed too low to cause serious problems, the effect of copper and nickel required further study. The interference by copper in the stripping determination of zinc was extensively investigated by Bradford (8). He concluded that in the mercury film, copper and zinc formed a 1:1 intermetallic compound that dissociated to release zinc during the oxidation. Thus zinc peak areas remained proportional to the zinc concentration even in the presence of copper, and the analysis of zinc by standard addition was not affected. The interference from nickel was found to be similar to that from copper although the stoichiometry of the intermetallic compound could not be determined.

To summarize, the analysis of seawater samples of representative composition for copper, zinc, cadmium, and lead by AASV with standard addition should yield reasonably accurate values for the concentrations of the metals. Although nickel and silver are present in seawater in concentrations high enough to interfere with the determinations of zinc and copper, the error caused by these metals is expected not to exceed 10 or 15%. Nevertheless, the composition of samples cannot always be guaranteed, and the analysis is always made on the assumption that the standard partitions are present in the sample and in the film in exactly the same manner as the metals originally present in the sample. Because this cannot be known with certainty, particularly when a field survey is being conducted, automated ASV with the thin film must at present be considered a semiquantitative indicator of trace metal activity in the water. Thorough intercomparisons between thin-film voltammetry and other techniques are needed to establish fully the quantitative aspects of this method.

Initial efforts in this direction have produced mixed results. In a careful study, Huynh Ngoc (9) compared zinc concentrations obtained by ASV using a mercury-covered graphite electrode with concentrations obtained by extraction followed by visible spectrophotometry and by atomic absorption spectrophotometry. Extraction yields were determined radiometrically. Atlantic Ocean samples were determined before and after strong oxidation procedures, and six separate determinations were performed with each method. For the untreated samples the mean result obtained by ASV was within 10% of the mean result obtained by the other methods. For the acidified, oxidized samples, ASV yielded a mean value within 20% of that produced by spectrophotometry.

A multilaboratory program is in progress to compare lead values obtained by ASV with the thin film with those determined by isotope dilution. Initial intercomparisons with samples of Southern California coastal waters were disheartening but were tentatively traced to procedural errors and contamination. Recent efforts have produced some values which are within a factor of two of the standardized lead value (*10*). Our own initial efforts to compare copper concentrations obtained using the system described herein with those obtained by extraction on Chelex 100 and subsequent analysis by atomic absorption showed ASV values to be considerably higher. This discrepancy was traced to interference by silver from the reference electrode. Recent work with a noncontaminating reference electrode yielded copper values that were essentially identical to those obtained by extraction followed by atomic absorption (*11*).

Variations in the Coverage of the Mercury Film. During analysis, mercury coverage of the graphite tube is at steady state (*1*). This steady state depends on the duration of the mercury plate, the electrolysis potential, the concentration of the plating solution, and the flow rate and chemical nature of the seawater itself. Changes in the water caused, for instance, by varying the pH, as well as changes in the time of electrolysis, will lead to a new steady state if these parameters are then kept constant. Because of the large overvoltage required for the deposition of mercury (*12*), changes of the electrolysis potential will also disrupt the steady-state coverage. Thus, the electrode does not lend itself to experiments in which chemical and physical variables are rapidly changed. When such changes occur, the electrode must be allowed to pass through several plating–stripping cycles to re-equilibrate; otherwise the current–voltage data will lead to ambiguous results.

Linearity with Concentration. Automated ASV of zinc in Gulf of Mexico samples produced a standard addition curve which was linear in the 1×10^{-9}–$1 \times 10^{-8}M$ range. The linearity of zinc standard additions in the 2×10^{-8}–$2 \times 10^{-7}M$ range has already been shown (*1*). Lead and cadmium standard additions to Scripps Pier water were also linear in the 1–$5 \times 10^{-9}M$ range. Standard additions of copper to San Diego Bay water produced curves which showed a small negative deviation for linearity in the 1–$6 \times 10^{-8}M$ range. This curvature has been ignored in our calculation of copper concentrations. These ranges of linearity are only those studied, not maximum ranges.

Analytical Precision. At sea, replicate plating–stripping determinations of 100-ml zinc subsamples containing approximately $1 \times 10^{-8}M$ zinc yielded peak currents with a standard deviation (one sigma) of 3.1%. In the laboratory, a seawater sample containing approximately $1 \times 10^{-8}M$ copper and $1 \times 10^{-9}M$ lead was acidified to pH 1 with hydrochloric acid and measured by DPASV. The copper peak currents deviated 1.2% from the mean. Lead peak currents deviated 2.8% from the mean. Replicate plating–stripping determinations do not give a true measure of the analytical precision, however. That requires addition of a standard and full equilibration of the standard with the sample and the container and can be better computed from the slopes of standard addition plots. In the $1 \times 10^{-8}M$–$1 \times 10^{-7}M$ range these result in an error of 10% for zinc, 11% for cadmium, and 15% for lead. The considerable difference between replicate plating stripping measurements and the analytical precision again points to the important role of the standard in this determination.

Observations on the Pier of the Scripps Institution of Oceanography. In August 1973 the automated ASV system was placed at the end of the 300-ft Scripps pier over approximately 10–15 ft of water, depending on tidal conditions. Seawater was sampled with a 50-ft length of Tygon tubing placed several feet below the surface. The apparatus was operated continuously each working day (8 hr) for two weeks. Electrodes and mercuric nitrate solutions were freshly prepared each day. Analyses were performed continuously by linear-sweep ASV using the Heathkit instrument, and approximately every fifth sample was redetermined after standard addition. An electrolysis time of 12 min at -1.4 V *vs.* silver/silver chloride was chosen, and most samples were measured at a pH slightly above the natural pH after sparging with prepurified nitrogen. Some samples were measured at pH 4.9 by sparging with carbon dioxide.

Initially, it was observed that successive analyses were essentially identical and showed little variation with time. The concentration levels of zinc, lead, and "copper" (contaminated with silver) remained relatively consistent for six days, declined over a period of about a day and then assumed a new low level for one more week. The results of the observations made at the Scripps pier are summarized in Table I. Only those determinations made by standard addition are included in the table values. Thus, the data represent approximately 50 individual measurements. A study of Table I reveals that the measurements in the period 8-10 to 8-15 belong to a different statistical population than the measurements made during 8-17 to 8-24. Within each population each metal varied an average of ±33%—considerably more than the analytical error. Since the variability was virtually identical for each indicator metal, it may have been caused by temporal changes in the water.

Greater peak currents were observed for lead at pH 4.9 for both the sample and the added standard, but the calculated concentrations were the same as those observed at pH 8. This suggests that the standard was in equilibrium with the sample at each pH.

Table I.   Tabulation of Zinc, Lead, and "Copper" Concentrations for Scripps Pier Water

	Concentration ($\mu g/l.$)		
Date	*Zn*	*Pb*	*"Cu"*
8-10 to 8-15	2.30 ± 31%	0.44 ± 34%	0.69 ± 33%
	(n = 13)	(n = 13)	(n = 9)
8-17 to 8-24	0.44 ± 39%	0.20 ± 30%	not
	(n = 9)	(n = 12)	measurable
Overall mean	1.54	0.33	0.69

Observations in Puget Sound, Wash. The automated ASV system with the solid-state programmer was tested aboard the University of Washington research vessel Thomas G. Thompson. Samples of Puget Sound water were collected while underway from Seattle, Wash., to Saanich Inlet, B.C. After a four-day period in Saanich, the ship returned to Seattle over essentially the same course. Samples were collected underway by attaching the Tygon tubing to the boat boom and towing it at full speed (about 2 knots) approximately 100 ft behind the ship. The tubing was held about 1 ft below the surface by an iron chain fastened 10 ft from the tubing inlet. A flow of seawater was maintained in the ship's laboratory during the transit.

Analyses for "copper," cadmium, and lead were carried out continually by DPASV. Zinc determinations were excluded to permit use of a lower electrolysis potential. The samples were analyzed at pH 4.9 by sparging with carbon dioxide. An 8-min. electrolysis at −1.0 V *vs.* silver/silver chloride and a 25-mV pulse were used during the Seattle-Saanich portion of the trip (Leg 1) while a 10-min. electrolysis and a 50-mV pulse were used from Saanich to Seattle (Leg 2). Application of the DPASV technique resulted in greater sensitivity and thus shorter plating times for the low levels encountered. It also afforded better resolution for "copper" than linear-sweep ASV. It should be pointed out, however, that DPASV does not result in shorter analyses times because the stripping portion of the analysis is very slow. Nevertheless, it is worthwhile to limit the time of electrolysis because this also reduces the concentrations of interfering metals accumulated in the mercury film. Under the

Table II.   Tabulation of Cadmium and Lead Concentrations for Puget Sound Seawater

Date	Time (PDT)	Metal Concentrations (μg/l.)	
		Cd	Pb
	Leg 1—Seattle to Saanich		
8/26/74	2200	N.D.[a]	0.16
8/26/74	2330	0.1	0.14
8/27/74	0030	0.1	0.13
8/27/74	0050	0.1	0.11
8/27/74	0130	0.1	0.21
8/27/74	0150	0.1	0.13
8/27/74	0310	0.1	0.10
8/27/74	0330	0.1	0.11
8/27/74	0350	0.1	0.10
Mean—Leg 1		0.1	0.13
	Leg 2—Saanich to Seattle		
8/30/74	2316	N.D.[a]	0.06
8/30/74	2337	N.D.	0.03
8/31/74	0046	N.D.	0.03
8/31/74	0106	N.D.	0.02
8/31/74	0130	N.D.	0.05
8/31/74	0213	N.D.	0.07
8/31/74	0326	N.D.	0.03
Mean—Leg 2		N.D.	0.04

[a] N.D. indicates signal was not detectable over baseline.

operating conditions described above, a complete analysis could be carried out every 20 min.

Three types of determinations were made while underway: the sample was analyzed, the sample and a standard addition were analyzed, and an internal standard solution was measured. San Diego Bay water which had been passed through a Chelex 100 ion-exchange column was chosen as the internal standard. The standard appeared to have higher lead and approximately equal "copper" when compared with the Puget Sound samples.

Data obtained on Legs 1 and 2 are presented in Table II. Concentrations were calculated by comparison with an added standard which increased the concentrations of the samples by $2.0 \times 10^{-8}M$ cadmium and $2.0 \times 10^{-9}M$ lead. Results for measurements which were not followed by standard additions were estimated by interpolating between bracketing standard additions. The concentration levels are those generally expected for open-ocean water (13). Lead values are somewhat higher than those of Southern California surface waters recently measured by isotope dilution (10), but it must be noted that these data were not corrected for an analysis "blank," which in our system may include desorption from the cell walls and contamination from the mercuric nitrate solution. Principally, we have not been able to prepare a sufficiently clean salt solution to run through our cell to determine the "blank."

Interestingly, measurements made on Leg 2 yielded values a factor of three lower than those made on Leg 1 for each of the metals detected; this observation is similar to that made on Scripps pier. Again, the diminution of trace metals appears to be related to some change which has occurred in the water. Several reasons can be advanced to support this statement.

1. The measured concentrations appear to be independent of the quality of the performance of the electrode, as determined by standard additions.

2. Prior to our departure on Leg 1, Puget Sound experienced stormy, windy weather followed by a lengthy period of sunlight and warm weather which lasted until our return on Leg 2. Planktonic uptake occurring during the interval between Legs 1 and 2 could account for the diminution in trace metal concentration. This observation would then agree with an earlier observation made in the surface waters of the open sea (*14*).

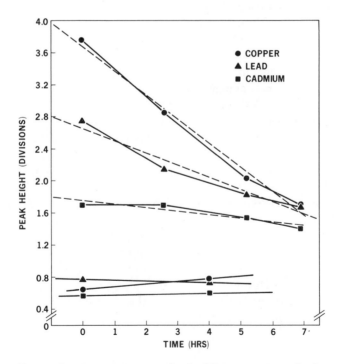

Figure 4.   Response to standard addition vs. time of addition for Puget Sound seawater. Lower plots, Leg 1. Upper plots, Leg 2.

3. Organic matter exuded by the plankters could have complexed with the metals and made them more difficult to reduce at the electrode. The presence of organically complexed trace metals in lake water has been shown (15, 16). If the added standard did not fully equilibrate with the organic matter, apparently low concentrations would have been obtained.

Another inference concerning differences in the water may be made by plotting the peak heights of the added standards vs. the time of the addition (Figure 4). Peak heights for Leg 2 were initially substantially higher than those for Leg 1 even when normalized for pulse amplitude (50 mV vs. 25 mV) and electrolysis times (10 min vs. 8 min). These peaks declined rapidly with time and eventually reached the same levels as the normalized standards in Leg 1. The standard additions performed during Leg 1 showed no significant changes in current during a 5-hr period. At present we can offer no explanation for this behavior other than to suggest that electrode sensitivity may have been reduced by trace substances, possibly of organic nature, retained by the electrode. Because each electrode was new at the outset of each leg and because each electrode was subjected to approximately the same number of determinations, it is unlikely that the sensitivity decrease was caused by the major inorganic constituents of the water.

An indication of the signals produced by the automated ASV system on Leg 1 is shown in Figure 5. The illustration displays a series of current–voltage scans made over a 3-hr period. The most cathodic peak is an unidentified contaminant present in the mercury solution. It does not appear to interfere with the determinations and serves to indicate electrode performance. The initial current–voltage scan (A) shows the trace obtained for the independent standard, the San Diego Bay water "purified" by ion exchange. The second scan (B) shows the current–voltage plot for the seawater sample collected from behind the ship. The third (C) illustrates a standard addition. Part of the added standard is carried over to analysis (D), which must then be disregarded. Current–voltage scans (E)–(G) are the results of line samples processed by the system. Finally, in (H), a second determination of the independent standard is shown which is identical to (A). The relative reproducibility of the system and the remarkable uniformity of the surface waters are notable. Because the ship was travelling at approximately 10 knots when these determinations were made, these samples are representative of approximatly 30–40 mi of water surface.

Conclusion

Automated ASV performed in the field provides useful data for the study of certain trace metals in the marine environment. The advantages of the system are numerous: rapid analyses can be performed on ship-board while underway; the precision of the method is good; sample handling is minimized, thereby eliminating contamination from sources outside the system; and the process lends itself to computerization. However, the accuracy of the method has not been conclusively established.

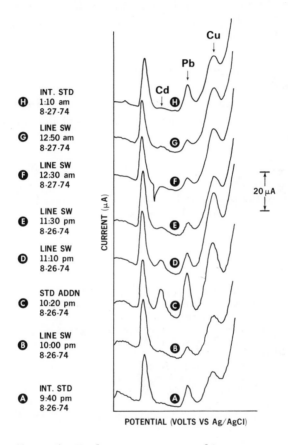

Figure 5.   Peak current vs. potential for consecutive shipboard analyses of Puget Sound seawater

Although a thorough understanding of the electrode processes is still lacking, changes in "apparent concentrations" may be used to study physical and biological processes occurring in the oceans. Thus, data obtained by ASV are not much different from data obtained by other methods which have to be interpreted in terms of an operational definition, such as "apparent" pH, "reactive" silicate, etc.

Finally, it must be pointed out that the automated ASV system presented herein is the result of an experimental concept which offers the opportunity for much variation and that the methodology we used is not necessarily optimum, but just a technique which produced data with some apparent analytical and oceanographic significance. Improvements are possible. For example, a modification of the procedure which might yield better results would be to replace the seawater during the stripping step with a medium in which the oxidation of the metals would be more nearly reversible (*17*). Medium exchange can be accomplished conveniently with the tubular electrode configuration and would produce better signals for more metals and reduce the level of interfering contaminants. Indeed, because of the great possibility for experimentation, the hard-wired programmer we described is already obsolete and will be replaced by a programmable minicomputer. It is expected that in the future such systems will contribute new insights of the behavior of trace metals in the oceans.

Acknowledgment

We wish to thank Sachio Yamamoto of the Naval Undersea Center's Chemistry and Environmental Sciences Division for his continuing support of this project.

Literature Cited

1. Lieberman, S. H., Zirino, A., *Anal. Chem.* (1974) **46**, 20.
2. Zirino, A., Healy, M. L., *Environ. Sci. Technol.* (1972) **6**, 243.
3. Morris, J., private communication (1974).
4. Smith, J. D., Redmond, J. D., *J. Electroanal. Chem.* (1971) **33**, 169.
5. Seitz, W. R., Ph.D. Thesis, The Massachusetts Institute of Technology, Cambridge, Mass., 1970.
6. Schutz, D. F., Turekian, K. K., *Geochim. Cosmochim. Acta* (1965) **29**, 259.
7. Spencer, D., Brewer, P., *Geochim. Cosmochim. Acta* (1969) **33**, 325.
8. Bradford, W. L., Ph.D. Thesis, The Johns Hopkins University, Baltimore, Md., 1972.
9. Huynh Ngoc, L., Ph.D. Thesis, Université de Nice, Nice, France, 1973.
10. Brewer, P., *et al.*, *Mar. Chem.* (1974) **2**, 69.
11. Zirino, A., *et al.*, in preparation.
12. Hume, D. N., Carter, J. N., *Chemia Analityczna* (1972) **17**, 747.
13. Chester, R., Stoner, J. H., *Mar. Chem.* (1974) **2**, 17.
14. Zirino, A., Healy, M. L., *Limnol. Oceanog.* (1971) **16**, 773.
15. Allen, H. E., Matson, W. R., Mancy, K. H., *J. Water Pollut. Control Fed.* (1970) **42**(4), 573.
16. Chau, Y. K., Lum-Shue-Chau, K., *Water Res.* (1974) **8**(6), 383.
17. Zieglerova, L., Stulik, K., Dolezal, J., *Talanta* (1971) **18**, 603.

RECEIVED January 3, 1975. This work was supported by the Defense Research Project Agency and by the Office of Naval Research under contract NR083-301.

9

Mercury Analyses in Seawater Using Cold-trap Pre-concentration and Gas Phase Detection

WILLIAM F. FITZGERALD

Marine Sciences Institute and Department of Geology,
University of Connecticut, Groton, Conn. 06340

A cold-trap pre-concentration procedure, which is incorporated into a standard flameless atomic absorption analysis of mercury in environmental samples, has been used for both shipboard and laboratory analyses of mercury in seawater. The coefficient of variation for seawater containing 25 ng Hg/l. is 15%, and a detection limit of approximately 0.2 ng Hg is attainable. In surface seawaters of coastal and open regions of the northwest Atlantic Ocean mercury concentrations appear to decrease with increasing distance from terrestrial sources. In the open ocean samples they are less than 10 ng/l. and rather uniformly distributed. The amounts of mercury in inshore samples can approach 50 ng/l. A significant mercury fraction characterized by a stable association with organic material may be present in coastal waters.

Among the preferred analytical methods for determining mercury concentrations in natural samples save been closed system reduction–aeration procedures using mercury detection by gas phase atomic absorption or atomic fluorescence spectrophotometry (*1–15*). In studies in the oceanic regime, where the amount of mercury in a liter sample of open-ocean seawater can be as small as 10 ng (*11, 15, 16, 17*), a pre-concentration stage may be required. The lowered detection limits which accompany a preliminary concentration step are most desirable when the sample materials are rare or in limited quantities such as carefully collected open-ocean biota, open-ocean rain water, and deep-ocean seawater.

Concentration of mercury prior to measurements can also separate mercury from interfering substances. Amalgamation with a noble metal (3, 10, 12, 15) and dithizone extraction (5, 14, 17) have been commonly used to both separate and concentrate mercury for analysis. I thought, however, that a simple condensation trap, similar to the collecting device used with volatile hydrocarbon determinations in seawater (18), would effectively concentrate mercury without contamination. Knudson and Christian (19) described a simple U-tube cold trap that was used to determine volatile hydrides of arsenic, antimony, bismuth, and selenium by flameless atomic absorption. In addition, procedures for determining mercury in geological (9) and biological materials (20) have used cold traps at the temperature of liquid nitrogen.

My students and I have carried out mercury analyses in seawater using a cold-trap pre-concentration modification of the closed system reduction–aeration flameless atomic absorption procedure described by Hatch and Ott (1). The cold trap is created by immersing a glass U-tube packed with glass beads in liquid nitrogen. After reduction with stannous chloride, purging, and trapping, the mercury is removed from the glass column by controlled heating, and the gas phase absorption of eluting mercury is measured. This procedure has been applied to both shipboard and laboratory analyses of mercury in seawater. A detection limit of 2 ng Hg/l. is attainable using a 100-ml seawater sample.

Careful and accurate sampling is an integral and important part of analytical procedures for measuring trace constituents in the marine environment. Thus, the sampling methodology used here for mercury measurements in seawater is discussed. In addition, representative results for the amounts and distribution of mercury in the northwest Atlantic are given. Finally, mercury measurements are presented which suggest the presence in coastal waters of very stable organo–mercury associations. The analytical pursuit of such chemical species of mercury can serve as one point of departure for further studies of mercury in the oceans.

Experimental

This section presents an overview of the cold-trap pre-concentration gas phase detection method for the determination of mercury in seawater. A more thorough and complete discussion of the analytical details (e.g., cleaning procedures, reagent preparation, analytical manipulations) can be found in Fitzgerald et al. (21).

Analytical Apparatus. In the cold-trap method, mercury is detected by the gas-phase absorption of elemental mercury at 2537 A using a Coleman Instruments mercury analyzer (MAS-50) equipped with a Leeds and Northrup Speedomax recorder (model XL601). The mercury analyzer is placed at one end of a sampling train with nitrogen as the purging and carrier gas. The gas-flow system also includes a 250-ml borosilicate glass (Corning Glass Works) bubbler, a flow regulator, two bypass valves, a water absorber, a mercury cold trap, a gas cell, and a gas washing bottle containing a 10% solution of potassium permanganate. A schematic of the analytical system appears in Figure 1.

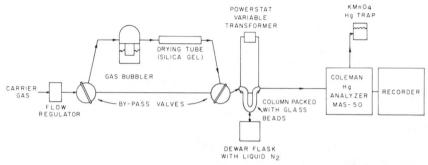

Analytical Chemistry

Figure 1.   Schematic of the mercury cold-trap pre-concentration system for mercury determination by gas phase absorption (21)

The seawater sample solution (100 ml) is contained in the 250-ml bubbler. In turn this sparging vessel is connected directly to a borosilicate glass drying tube containing colorless silica gel (6/20 mesh) as a water absorbent. The mercury cold trap follows this absorbing stage and consists of an 8-in. borosilicate glass (id 4 mm) U-tube (width 1.5 in.) packed with glass beads (80/100 mesh) to form a 6-cm column at the bend of the tube. The U-tube was wrapped with Nichrome wire (diameter 0.06 mm) yielding five windings/in. over the entire lower 6 in. of the tube. The wire-wound U-tube was placed inside an insulating borosilicate glass covering such that only the non-wire wrapped sections were exposed. This complete U-tube apparatus is designed to fit a 1-l. Dewar flask containing liquid nitrogen to provide the trap and concentration step for mercury vapor. The leads from the heating wire were connected to a Powerstat variable transformer which allowed the column to be heated electrically when the liquid nitrogen bath was removed.

Three-way Teflon (DuPont) stopcocks (4 mm bore) are placed before the bubbler and after the drying tube to permit the sparging vessel and the drying column to be bypassed during the heating and elution step. The mercury, which is rapidly vaporized and eluted from the column during the heating step, was fed directly to the gas cell of the Coleman mercury analyzer. The absorption of elemental mercury in arbitrary units was displayed on the recorder, using a 25× scale expansion.

After the carrier gas had passed through the gas cell, it was directed into a 300-ml gas washing bottle containing 100 ml of 10% potassium permanganate. At this stage, the elemental mercury was oxidized and removed from the gas flow.

Analytical Procedure. The cold-trap gas phase mercury detection system was designed and used for both laboratory and shipboard measurements of mercury in seawater. The Coleman Instruments mercury analyzer (MAS-50) was incorporated into the analytical system because of its portable and convenient design. However, the effective use of this simple one-element atomic absorption unit requires scrupulous attention to blank determinations for each seawater sample. For example, the undetected presence of either naturally occurring or sampling induced volatile organics which may absorb at the mercury wavelength in the seawater sample can be a serious error. Such artifacts were observed when acidified seawater samples were stored in low density polyethylene bottles (21). Therefore, the analytical procedure used to determine the mercury concentration in a seawater sample consists of the following steps:

1. An initial blank determination for each seawater sample

2. The stannous chloride reduction–nitrogen purging step followed by the gas-phase measurement of elemental mercury

3. The determination of the mercury response associated with standard mercury spikes added to the sample matrix

4. A repeat of the blank determination with the seawater sample now containing stannous chloride but devoid of mercury.

Sample Determination. With the U-tube column immersed in the liquid nitrogen bath, a 100-ml seawater solution (pre-acidified to ~ pH 1 with the high purity nitric acid) is placed in the gas bubbler. A 0.5 ml addition of the stannous chloride reagent (20%) reduces the mercury in the sample to its elemental state. The sparger is inserted, and the seawater solution is mixed by hand shaking for 5 sec, and then the Teflon stopcocks are switched to the flow-through position. The purging rate for the nitrogen aeration is 0.5 l./min at 7 psi. After purging is completed (7 min), the valves are returned to the bypass position, and the flow rate of the carrier gas is increased to 0.7 l./min (7 psi).

The column is removed from the liquid nitrogen bath, and the U-tube is heated through the wire windings using the variable transformer. The transformer can be simply switched on by prior calibration of the voltage setting to give a column temperature of 225°C measured on the outside wall of the U-tube after 60 sec of heating. The elemental mercury is vaporized and eluted from the column in 1–2 sec, and the entire operation requires less than 10 sec from the time the U-tube is removed from the cold trap. During the heating and elution cycle (60 sec), the flow rate of the carrier gas decreases from 0.7 l./min to 0.15 l./min.

The absorption at the mercury wavelength (2537 A) occurs in the gas cell 12 sec after the heating cycle is initiated. The absorption is recorded in arbitrary units, and the maximum height is noted. After the response has returned to the initial base line (60 sec), the heating is stopped, and the column is cooled in air for 30 sec. The U-tube is then returned to the liquid nitrogen bath. The carrier gas flow is set to 0.5 l./min, and the system is ready for the mercury spike additions. Total time for this operation is 9 min. Except for the stannous chloride addition, the system blank is established in a manner identical to the sample determination.

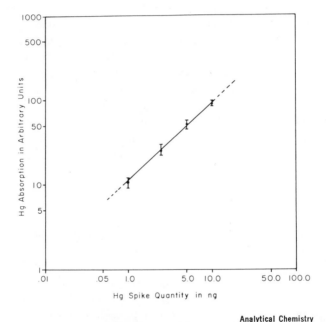

Analytical Chemistry

Figure 2. A composite calibration curve for mercury in seawater samples measured over a three-week period. The mercury spike quantities are in ng and the mercury absorption in arbitrary units (21).

Analytical Curves. Appropriate spikes of mercuric chloride standards were added to the sample matrix, and the sample determination procedure was repeated. Three spike additions were usually made. A sample blank is measured and the mercury concentration determined from an individual calibration curve for each sample. The working calibration curve for a seawater sample is prepared by plotting the gas phase absorption of mercury (arbitrary units) against the mercury spike concentration (ng/l.). Using log–log coordinates, the mercury response is linear over a wide concentration range (10–200 ng Hg/l.), which brackets conveniently the expected variability for mercury in marine waters.

In Figure 2, a composite calibration curve for seawater is reproduced. This curve is based on the response obtained for spike additions of 1.0, 2.5, 5.0, and 10 ng Hg to different 100-ml seawater samples analyzed over three weeks. The number of spikes making up this composite graph are 3, 19, 20, and 9, for the 1.0-, 2.5-, 5.0-, and 10-ng mercury additions, respectively. The average value for each addition was plotted, and the brackets indicate the standard deviation. The precision of analysis reported as a coefficient of variation for these spike additions is 30% at 5 ng Hg/l., 20% at 10 ng Hg/l., 15% at 25 ng Hg/l., and 10% at 50 ng Hg/l.

This composite calibration curve for seawater demonstrates the applicability of the cold-trap pre-concentration technique to low concentration ranges of mercury. Approximately 0.2 ng of mercury can be determined with a $25\times$ scale expansion. Since the response depends on the vaporization and elution of trapped mercury from the column, the calibration curves were similar for other aqueous media including acidified (nitric acid) distilled deionized water. Therefore, this cold-trap procedure appears to separate effectively reducible mercury species from interfering substances that might be associated with differing solution matrices.

Field Investigations

The accurate determination in seawater of trace constituents, such as mercury, requires not only an accurate analytical method to measure the constituent but scrupulous care to prevent contamination during the shipboard sampling and sample manipulations. A sampler and a sampling procedure (e.g., shipboard handling) must be used that do not affect the true concentration of the constituent either during collection or during the period awaiting transfer to a storage container. Moreover, a sample storage container and preservation technique must be used that do not alter or artificially affect the amounts of the constituent in the stored seawater prior to analysis.

Some of the potential problems that may affect mercury determinations in seawater are listed below.

1. Adsorption losses to the sample container during storage (22, 23, 24). These investigations suggest that under nonacidified conditions, mercury adsorption to the container walls is rapid enough to affect mercury concentrations within a sampler during a hydrocast.

2. Volatile organics, either naturally occurring or sampler-induced, may interfere with adsorption at the mercury wavelength (2537 A) during the usual flameless atomic absorption procedure (4).

3. The mercury concentration in the seawater sample may be deleteriously affected by the sampler or the shipboard manipulations.

4. The analytical method may not measure all the chemical fractions of mercury present (*25*).

We found that seawater solutions pre-acidified with nitric acid to 1.2–1.3% by volume and containing 5–50 ng Hg/l. could be stored in Teflon bottles for at least one month without significant loss of mercury. Therefore, we used Teflon bottles for storage containers and collection samplers for surface seawater. We found borosilicate glass containers to yield occasional high blanks. Moreover, the presence of either volatile organic plasticizer material or polyethylene residue leached by the acidified seawater solution renders polyethylene containers unsuitable.

Open Ocean Mercury Determinations. In our initial studies concerned with the marine geochemistry of mercury, we obtained open ocean surface samples by hand from a small work boat away from any adverse influence of the oceanographic research vessel. The concentrations of mercury in the open-ocean surface waters (western Sargasso Sea) were small (*ca.* 10 ng/l.) and rather uniformly distributed (*26*). However, to collect seawater to determine the concentrations of mercury at other depths, we needed an artifact-free sampling procedure.

Table I.   A Comparison of Mercury Determinations in Surface Seawater Collections from Pre-acidified Teflon Bottles and from 5-l. PVC Samplers[a]

Station Number	Station Location	Salinity (‰)	Temperature (°C)	Pre-acidified Teflon Bottle	5-l. PVC Sampler[b]
1	39°56.6'N 66°18.2'W	35.31	14.0	6	8
2	36°35.3'N 66°03.2'W	36.26	23.8	6	5
3	32°54.4'N 66°07.0'W	36.57	21.1	10	9
4	32°03.0'N 64°54.0'W	36.10	22.5	8	10

Mean and standard deviation of the 24 analyses—8 ± 3
Blank determinations (Stations 2–4)—below detection limits

[a] Trident cruise 152, May 1974.
[b] Manufactured by General Oceanics, Inc., Miami, Fla.

We devised the following scheme to test the suitability of a widely used PVC sampler (General Oceanics, Inc., Miami, Fla.) with a Teflon coated stainless steel closing spring. During R/V Trident cruise 152 (May 8–19, 1974), we obtained both surface seawater samples and collections at depth for vertical mercury profiles (to 750 m) at four stations between Narragansett, R.I. and Bermuda. The locations are indicated in Table I. Prior to each hydrocast, surface-water samples were collected from a rubber work boat away from the influence of the Trident. The workboat was powered by a small electric motor, and the surface collections were carefully made off the bow while underway. These surface samples were collected by hand, using polyethylene gloves, in Teflon bottles pre-acidified with concentrated nitric acid (J. T. Baker Chemical Co., Ultrex) to yield 1.2–1.3% nitric acid in the seawater sample. Following the surface seawater collection in Teflon, a sample of the surface seawater was also captured in a 5-l. PVC sampler. The Teflon bottles and the sampler were stored before and after collection in a polyethylene bag to avoid contact with the rubber work boat and the ship.

Immediately upon return to the ship, the hydrocast was initiated with the closed PVC sampler (containing the surface water collection) placed as the first bottle on the hydrowire followed by four open and cocked samplers set at various depths from 25 to 750 m. The hydrocast and sample handling were then carried out in the usual fashion.

Table II.   Mercury Concentrations *vs.* Depth for Four Stations in the Northwest Atlantic Ocean[a]

	Hg Concentration (ng/l.)			
Depth (m)	*Station 1 (39°56.6′N 66°18.2′W)*	*Station 2 (36°35.3′N 66°03.2′W)*	*Station 3 (32°54.4′N 66°07.0′W)*	*Station 4 (32°03.0′N 64°54.0′W)*
0	7	6	10	9[11]
25	5[6]			
100		9[12]	8	4
250	7	10	8[5]	10
500	6	12	12	12[9]
750	8	10[10]	10	9

[a] Trident cruise 152, May 1974. Total mercury measurements are in brackets.

The results of this sampler study are summarized in Table I, where the value for the mercury concentration in each case is the mean mercury concentration found by triplicate analysis. The mercury concentrations found in the surface seawater collected in Teflon agree within experimental error with those for this same water after 1 hr in the PVC sampler subjected to the hydrocast procedure and sample transfer usually used. The mercury concentrations found at other depths (Table II) show little variation from the surface values. This suggests that the open samplers are not contaminated or affected significantly by the hydrocast procedure. Additional details regarding this sampling study can be found in Fitzgerald and Lyons (*27*).

The reported concentrations for mercury in seawater range from non-detectable to 364 ng/l. (*11, 12, 15–17, 28–30*). In several of these studies, the concentrations of mercury were greater than 100 ng/l. in open ocean waters (*12, 28, 29, 30*). The data from our investigations of the northwest Atlantic Ocean (Table II) and other studies (*11, 15, 16, 17*) indicate that the mercury concentrations should be closer to 10 ng/l. in these open ocean waters.

This variability for the reported concentrations of mercury in open ocean waters may indicate that there are significant analytical difficulties associated with the proper sample collection and the accurate measurement of mercury in seawater. These problems tend to override and preclude precise geochemical calculations and marine geochemical interpretations regarding the sources, sinks, and interactions of mercury in the oceans. These observational discrepancies for trace seawater constituents such as mercury can be resolved only through intercalibration programs and the use of standardized seawater samples. Such standards are not presently available for mercury concentrations at 100 ng Hg/l. or less in seawater.

Organo–Mercury Associations. Experimentally, the mercury measurements in seawater have been divided into two fractions—reactive and total mercury. The reactive fraction represents the amount of mercury measured in pre-acidified raw seawater samples at approximately pH 1. The total mercury measurement is carried out on aliquots of the pre-acidified seawater samples in which the organic matter has been destroyed by ultraviolet photooxidation (*31*). This irradiation procedure is as effective as the persulfate oxidation method (*32*) commonly used to destroy organic matter in seawater. A complete discussion of our photooxidation methodology can be found in Fitzgerald (*33*). The amount of mercury determined as the difference between the "reactive" and "total mercury measurements" represents a very stable organo–mercury association.

In our previous investigations of the amounts and distribution of mercury in the surface waters of the northwest Atlantic Ocean, we found a mean total mercury concentration of 7 ng/l. and a range of 6–11 ng/l. (*26*). Also, we found in open ocean surface waters no significant difference between the mercury concentrations measured directly in pre-acidified seawater ("reactive" mercury) and the total mercury determination in the "organic free" samples. In the work shown in Table II, we also found no significant difference between the reactive mercury determination and the "total mercury" measurement, which was carried out in approximately one third of the samples. The "total mercury" measurements appear in the square brackets for the results tabulated in Table II.

The mercury concentrations increase toward the continental shelf of the northeastern U.S., and chemical species of mercury characterized by a strong association with organic material appear to be present (26). In Table III, measurements that we have obtained for reactive and total mercury are summarized for the coastal waters of the New York Bight, Block Island Sound, and Georges Bank. The concentrations of total mercury found for the first two locations are not exceptionally large (27–45 ng Hg/l.). However, greater than 50% of this mercury can be associated strongly with organic material. This organic fraction, however, was not always evident. For example, the Georges Bank stations did not reveal this organo–mercury fraction.

Table III. Mercury Concentrations in Coastal Waters off the Northeastern United States

Oceanic Region	Time	Location	Depth	Hg Concentration (ng/l) Reactive	Total
N. Y. Bight acid dumping grounds	Oct 1973	40°22.0′N 73°34.7′W	13m	11	37
			23m	10	37
		40°19.2′N 73°32.8′W	8m	15	27
			18m	9	28
control		40°21.0′N 73°09.8′W	13m	12	30
Block Island Sound	Oct 1973	41°15.0′N 71°30.0′W	10m	33	45
Georges Bank	May 1974	40°52.0′N 70°18.1′W	~15cm	8	10
		42°20.5′N 67°13.8′W	~15cm	5	11
		41°12.4′N 66°30.7′W	~15cm	5	9

At present, we do not know how mercury is bound in the chemical species making up the organic mercury fraction. Moreover, our understanding of the role of organo–mercury species in the marine geochemistry of mercury is very limited and speculative. For example, the isolation and identification of organic mercury chemical species may provide a very useful means of tracing certain parts of the mercury cycle in the oceans. The analytical pursuit of organo–metallo species in seawater containing a myriad of unidentified organic material is quite a formidable although worthy endeavor.

Acknowledgment

The assistance of C. D. Hunt and W. B. Lyons in developing and applying the cold-trap method is gratefully acknowledged. Seawater sampling was conducted with the assistance of the captain and crew of the R/V Trident and colleagues from the Graduate School of Oceanography, University of Rhode Island.

Literature Cited

1. Hatch, W. R., Ott, W. L., *Anal. Chem.* (1968) **40**, 2085.
2. Hinkle, M. E., Learned, R. E., *U.S. Geol. Surv. Prof. Pap.* (1969) **650-D**, 251.
3. Fishman, M. J., *Anal. Chem.* (1970) **42**, 1462.
4. Lindsteat, G., *Analyst* (1970) **95**, 264.
5. Chau, Y-K., Saitoh, H., *Environ. Sci. Technol.* (1970) **4**, 839.
6. Omang, S. H., *Anal. Chim. Acta* (1971) **53**, 415.
7. Omang, S. H., Paus, P. E., *Anal. Chim. Acta* (1971) **56**, 393.
8. Hwang, J. H., Ullucci, P. A., Malenfant, A. L., *Cand. Spectros.* (1971) **16**, 1.
9. Aston, S. R., Riley, J. P., *Anal. Chim. Acta* (1972) **59**, 349.
10. Muscat, V. I., Vickers, T. J., Andren, A., *Anal. Chem.* (1972) **44**, 218.
11. Topping, G., Pirie, J. M., *Anal. Chim. Acta* (1972) **62**, 200.
12. Carr, R. A., Hoover, J. B., Wilkniss, P. E., *Deep-Sea Res.* (1972) **19**, 747.
13. Omang, S. H., *Anal. Chim. Acta* (1973) **63**, 247.
14. Gardner, D., Riley, J. P., *Nature* (1973) **241**, 526.
15. Olafsson, J., *Anal. Chim. Acta* (1974) **68**, 207.
16. Leatherland, T. M., Burton, J. D., McCartney, M. J., Culkin, F., *Nature* (1971) **232**, 112.
17. Chester, R., Gardner, D., Riley, J. P., Stoner, J., *Mar. Pollut. Bull.* (1973) **4**, 28.
18. Swinnerton, J. W., Linnenbom, V. J., *J. Gas Chromatogr.* (1967) **5**, 570.
19. Knudson, E. J., Christian, G. D., *Anal. Lett.* (1973) **6**, 1039.
20. Rook, H. L., Gills, T. E., LaFleur, P. D., *Anal. Chem.* (1972) **44**, 1114.
21. Fitzgerald, W. F., Lyons, W. B., Hunt, C. D., *Anal. Chem.* (1974) **46**, 1882.
22. Coyne, R. V., Collins, J. A., *Anal. Chem.* (1972) **44**, 1093.
23. Newton, D. W., Ellis, R. Jr., *J. Environ. Qual.* (1974) **3**, 20.
24. Feldman, C., *Anal. Chem.* (1974) **46**, 99.
25. Fitzgerald, W. F., Lyons, W. B., *Nature* (1973) **242**, 452.
26. Fitzgerald, W. F., Hunt, C. D., *J. Rech. Atmos.* (1974) **8**, 629.
27. Fitzgerald, W. F., Lyons, W. B., *Limnol. Oceanogr.* (1975) **20**, 468.
28. Robertson, D. E., Rancitelli, L. A., Langford, J. C., Perkins, R. W., "Baseline Studies of Pollutants in the Marine Environment (Heavy Metals, Halogenated Hydrocarbons and Petroleum)," background papers for a workshop sponsored by the N.S.F. Office for the I.D.O.E., Brookhaven Natl. Lab., 24-26 May, p. 231, 1972.
29. Weiss, H. V., Yamamoto, S., Crozier, T. E., Mathewson, J. H., *Environ. Sci. Technol.* (1972) **6**, 644.
30. Fitzgerald, R. A., Gordon, D. C., Jr., Cranston, R. E., *Deep-Sea Res.* (1974) **21**, 139.
31. Armstrong, F. A. J., Williams, P. M., Strickland, J. D. H., *Nature* (1966) **211**, 481.
32. Menzel, D. W., Vaccaro, R. F., *Limnol. Oceanogr.* (1964) **9**, 138.
33. Fitzgerald, W. F., Ph.D. Thesis, Dept. of Earth and Planet. Sci., M.I.T., and Dept. of Chemistry, W.H.O.I., 1970.

RECEIVED January 3, 1975. This work was supported in part by the National Science Foundation Office for the International Decade of Ocean Exploration Grant No. GX-33777 and by the Office of Sea Grant Programs, National Oceanic and Atmospheric Administration.

10

Acid–Base Measurements in Seawater

ROGER G. BATES and J. B. MACASKILL

Department of Chemistry, University of Florida, Gainesville, Fla. 32611

Acidity scales are used commonly to assess the chemical and biological state of seawater. The international operational scale of pH fulfills the primary requirement of reproducibility and leads to useful equilbrium data. Nevertheless, these pH numbers do not have a well defined meaning in respect to all marine processes. Seawater of 35‰ salinity behaves as a "constant ionic medium," effectively stabilizing both the activity coefficients and the liquid junction potential. It may be possible, therefore, to determine hydrogen ion concentrations in seawater experimentally. One method is based on cells without a liquid junction and is used to establish standard values of hydrogen ion concentration (expressed as moles of H^+/kg of seawater) for reference buffer solutions.

The sea is a living system and, like other living systems, its properties are a complex function of many chemical and biological processes. Some of these involve, directly or indirectly, the protonation of basic species, and consequently the state of the seawater system—its equilibrium processes and the rate at which equilibrium is being approached—depends on pH. Interactions within the hydrosphere, in which carbonate, phosphate, and silicate play an important role, regulate the pH within rather narrow limits, as the acid–base balance of the human body controls the pH of human blood.

The pH accordingly acquires primary importance as an index of the state of the many interactive acid–base systems of which seawater is composed. Two factors are essential to the most fruitful application of this acid–base parameter. The first is that pH measurements possess reproducibility of a high order, and the second is that the numbers obtained have a clear meaning in terms of the processes of interest. To achieve the necessary reproducibility, uniform procedures and standards for the measurement must be accepted by all workers in the field.

The pH scale has been defined operationally, and standard reference solutions based on a conventional scale of hydrogen ion activity have been selected (*1, 2*). Measurements of the pH of seawater made with different electrodes and instruments are satisfactorily reproducible when standardized in the same way (*3*). The results obtained, however, do not always have a clear interpretation. Formally, this difficulty can be attributed to the residual liquid junction potential involved in the measurement. The primary standards are necessarily dilute buffer solutions (ionic strength, $I \leqslant 0.1$) whereas seawater normally has an ionic strength exceeding 0.6. This difference in the concentrations and mobilities of the ions coming in contact with the concentrated solution of potassium chloride of which the salt bridge–liquid junction is composed gives rise to a potential difference that is indeterminate. Consequently, the measured pH is in error by an unknown amount and does not fall exactly on the scale fixed by the primary standards.

Pytkowicz and his co-workers (*4, 5, 6*) have avoided these difficulties in a manner that has proved highly suitable in evaluating dissociation equilibria in seawater. Apparent equilibrium constants are defined in terms of the operational pH value, which may be regarded as an apparent conventional hydrogen activity. Although the fundamental meaning of these two quantities is unclear in seawater, their combination yields correctly the desired ratio of acid to conjugate base, the correct solubility, etc. (*7*).

In another approach, Hansson (*8*) defines an activity scale based on a standard state in seawater rather than in the pure aqueous solvent. Standard values of pH for buffer solutions containing tris(hydroxymethyl)aminomethane (tris) in seawater of various salinities were obtained by a potentiometric titration procedure in a cell with glass and silver–silver chloride electrodes, essentially free from a liquid junction. Hansson's standard pH should correspond closely to $-\log m_H$ in the seawater system and accordingly can be compared with pm_H derived in the manner described later. In a manuscript kindly furnished by Dyrssen before publication (*9*), this scale was recently applied to *in situ* pH measurements.

This chapter examines some possible ways in which a pH scale which meets fully the requirements of oceanographic investigations might be established. First, the nature of the operational pH scale that has received international adoption (*2*) is reviewed.

The Conventional pH

The pH unit as formally defined by Sørensen in 1909 represented $\log (1/c_H)$, where c_H is the concentration of hydrogen ion in mol/l.

The experimental method by which Sørensen proposed to measure pH did not, however, actually provide an unequivocal value for the hydrogen ion concentration in solutions of unknown composition. Introduction of the concept of activity (a) and the activity coefficient (y, concentration scales or γ, molality scale) led to a modified definition (10) for which a modified symbol pa_H was first suggested:

$$pa_H = \log (1/a_H) \tag{1}$$

All difficulties were not resolved by this suggestion, however, as the activity coefficient of hydrogen ion could not be evaluated either experimentally or theoretically, independently of that of other counter ions present.

Most pH determinations are made by electrometric methods, the pH of the unkown solution (X) being calculated from that of a known standard (S) and the emf (E_X and E_S) of a cell composed of a hydrogen ion-responsive electrode (for example, a glass electrode or a hydrogen gas electrode) coupled with a reference electrode (calomel, silver–silver chloride). This cell is filled successively with the standard solution S and with the unknown solution X. A liquid junction potential E_j exists where these solutions make contact with the concentrated KCl solution of the reference electrode. From the Nernst equation for the cell reactions and assuming an ideal hydrogen ion response:

$$pH = pH(S) + \frac{(E_X - E_S)F}{RT \ln 10} + \frac{(E_{jS} - E_{jX})F}{RT \ln 10} \tag{2}$$

The operational definition was formulated by omitting the last term of Equation 2, that is, $\overline{E}_j$, the residual liquid junction potential expressed in pH units:

$$pH(X) = pH(S) + \frac{(E_X - E_S)F}{RT \ln 10} \tag{3}$$

Values of $(RT \ln 10)/F$ are listed elsewhere (11).

Under certain conditions regarded as ideal, $\overline{E}_j$ is probably actually close to zero. This should be the case when solution X matches closely the primary standard solutions S in pH, composition, and ionic strength (which must not exceed 0.1). Then pH(X) doubtless approaches $-\log (m_H \gamma_H)$, where m_H is the molality (mol/kg of water) of hydrogen ion and γ_H is its conventional activity coefficient on the numerical scale defined by the convention adopted for the assignment of values to pH(S). When, as in seawater, these conditions do not prevail, the meaning of the experimental pH(X) in terms of concentrations and activities becomes unclear.

pH and Hydrogen Ion Concentration

The determination of pH(X) in seawater is thus straightforward, but the interpretation of the numbers obtained leaves something to be desired. Furthermore, the transfer of glass electrodes from dilute buffer solutions to seawater may sometimes be accompanied by drifting potentials, caused by a changing asymmetry potential associated with the appreciable difference in the water activity of the two media. For this reason, it may be a sound procedure to use secondary standards composed of suitable buffers in artificial or natural seawater. A secondary standard consisting of artificial seawater buffered with tris was suggested by Smith and Hood (*12*). These reference solutions should have pH values near that of seawater itself (pH 8–9) and be highly stable. Standard pH values are assigned to them by careful comparison with the dilute primary standards, preferably with the use of a hydrogen gas electrode in a cell such as that shown in Figure 1.

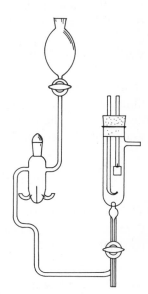

Figure 1.   pH cell with liquid junction. Calomel reference electrode on the left, hydrogen electrode compartment on the right. A liquid junction is formed in the capillary tube below the hydrogen electrode.

This procedure has certain experimental advantages, but it obviously cannot serve to clarify the meaning of the pH numbers obtained for seawater. Any pH value based on activities is subject, in many of its useful practical applications, to the uncertainties associated with the role of single ion activities in chemical and biological processes. By contrast, the hydrogen ion concentration is a well defined concept. In general, however, it is not feasible to derive $-\log m_H$ (pm_H) from pH(X) by estimating ionic activity coefficients, except at ionic strengths much lower than that of seawater. The pertinent relationships are made clear by comparing Equation 3 with Equation 2. Bearing in mind that pH(X) would be the "true" pH, that is, $-\log(m_H\gamma_H)$, if $\overline{E}_j$ could be evaluated, one has:

$$pm_H = pH(X) + \log \gamma_H + \overline{E}_j \tag{4}$$

and also

$$pm_H = pH(S) + \log \gamma_H + \overline{E}_j + \frac{(E_X - E_S)F}{RT \ln 10} \tag{5}$$

Constant Ionic Media

Seawater is a special type of solvent medium. Not only is its pH regulated within narrow limits, but its ionic strength is high, resulting largely from a single completely dissociated electrolyte (sodium chloride) while the molar ratios of sodium/chloride and magnesium/chloride are remarkably constant. Chemical and biological processes produce only small changes in composition compared with the total sea salt concentration. Seawater is accordingly a constant ionic medium which may be expected to "swamp out" changes in γ_H as these processes occur (13). If, furthermore, pH standards in seawater of the same salinity as the unknowns were available, $\overline{E}_j$ would be expected to drop to a value near zero. If these conditions were achieved, an experimental scale of pm_H could be set up. From Equation 5:

$$(pm_H)_X = (pm_H)_S + \frac{(E_X - E_S)F}{RT \ln 10} \tag{6}$$

The analogy with the operational definition of pH (Equation 3) is evident. It remains to explore the effectiveness of seawater in rendering γ_H constant and in nullifying the residual liquid junction potential, $\overline{E}_j$.

Activity Coefficients

To examine these questions, we have made emf measurements of cells of two types, namely:

$$\begin{array}{c|c|c}
\text{Pt;H}_2\text{(g,1 atm)} & \text{solution in} & \text{AgCl;Ag} \\
\text{or glass electrode} & \text{seawater} &
\end{array} \tag{A}$$

and

$$\begin{array}{c|c|c}
\text{Pt;H}_2\text{(g,1 atm)} & \text{solution in} & \text{3.5}M \text{ KCl,Hg}_2\text{Cl}_2\text{;Hg} \\
\text{or glass electrode} & \text{seawater} &
\end{array} \tag{B}$$

Table I.   Compositions of Two Types of Artificial Seawater

Seawater I (molality)	Constituent	Seawater II (molality)
0.4315	NaCl	0.4186
0.0531	MgCl$_2$	0.0596
0.0442	NaClO$_4$	0
0	Na$_2$SO$_4$	0.02856
0.01	KCl	0.01
0.005	CaCl$_2$	0.005
0.4757	Na$^+$	0.4757
0.5577	Cl$^-$	0.5577
0.553	Na$^+$/Cl$^{-\,a}$	0.553
0.0653	Mg^{2+}/Cl$^{-\,a}$	0.0733
19.1	chlorinity, ‰	19.1
35.6	salinity, ‰	34.2
0.66	ionic strength (I)	0.66

[a] Mass ratios

Cell A is a cell without liquid junction while cell B, with a liquid junction, resembles the cell assembly of the common pH meter.

The solvents consisted of artificial seawater of two compositions, with nearly identical salinities (35‰), chlorinities (19.1‰), and molalities of sodium and chloride ions (*see* Table I). The composition was close to that selected by Lietzke *et al.* (*14*). Sulfate was omitted from seawater I to avoid the complications of HSO$_4^-$ formation when strong acid was added. The ionic strength of the sulfate-free seawater was main-

Table II.   Electromotive Force of Cell A from 5° to 35°C: Solutions in Seawater I

Solute (Molality)	E/V			
	5°C	15°C	25°C	35°C
HCl, 0.01	0.37287	0.37259	0.37180	0.37054
0.03	0.34649	0.34527	0.34353	0.34131
0.05	0.33418	0.33253	0.33037	0.32774
HAc/NaAc, 0.01[a]	0.51048	0.51376	0.51694	0.52007
0.025	0.51034	0.51363	0.51682	0.51995
0.0442	0.51031	0.51360	0.51680	0.51990
Bis-tris/bis-tris HCl, 0.02[a]	0.64491	0.64367	0.64210	0.63995
0.04	0.64782	0.64628	0.64445	0.64205
0.06	0.64890	0.64721	0.64542	0.64294
Tris/tris HCl, 0.02[a]	0.74820	0.74317	0.73789	0.73234
0.04	0.74835	0.74333	0.73805	0.73253
0.06	0.74850	0.74348	0.73822	0.73271

[a] Molality of each component of the buffer

tained at 0.66 by adding sodium perchlorate instead of sulfate. When sulfate was present, the stoichiometric molalities were adjusted, taking account of ion pairing (15), to provide the desired ionic strength. The compositions in the table are expressed as molalities, that is, mol/kg of water.

Five series of solutions were prepared in artificial seawater. The solutions were hydrochloric acid, an acetate buffer solution (m_{HAc}/m_{NaAc} = 1), and three equimolal buffer solutions (m_{BHCl}/m_B = 1) prepared from the following bases (B): tris, 2-amino-2-methylpropanediol (bis), and the N-bis(hydroxyethyl) derivative of tris, bis-tris. The pK_a values of the protonated bases in water at 25°C are, respectively, 8.075 (16), 8.801 (17), and 6.483 (18). When hydrochloric acid or buffer was added to the seawater solvent, the ionic strength and chloride molality were kept constant by reducing the molalities of sodium chloride or sodium perchlorate as necessary.

Measurements of the emf (E) of cell A were made at 5°, 15°, 25°, and 35°C, and the data are summarized in Tables II and III. The results for solutions of hydrochloric acid were used to calculate the mean activity coefficient $\gamma_\pm$ of hydrochloric acid at three low molàlities (m) in artificial seawater (without sulfate):

$$- \ln \gamma_\pm = \frac{(E - E°)F}{2RT} + \frac{1}{2} \ln (mm_{Cl}) \tag{7}$$

The values plotted in Figure 2 display the expected linear variation with molality of hydrochloric acid at constant ionic strength. The intercept measures the trace activity coefficient, γ_{HCl}^{tr}, the limit of $\gamma_\pm$ in pure (acid-free) seawater. At 25°C, γ_{HCl}^{tr} = 0.731 as compared with 0.728 in

Table III. Electromotive Force of Cell A from 5° to 35°C: Solutions in Seawater II

Solute (Molality)	E/V			
	5°C	15°C	25°C	35°C
Bis-tris/bis-tris HCl, 0.02[a]	0.64521	0.64404	0.64242	0.64031
0.04	0.64802	0.64660	0.64469	0.64229
0.06	0.64903	0.64756	0.64558	0.64310
Tris/tris HCl, 0.02[a]	0.74920	0.74423	0.73900	0.73349
0.04	0.74928	0.74430	0.73907	0.73356
0.06	0.74930	0.74434	0.73912	0.73362
Bis/bis HCl, 0.02[a]	0.79129	0.78703	0.78241	0.77752
0.04	—	—	0.78255	—
0.06	0.79146	0.78721	0.78261	0.77767

[a] Molality of each component of the buffer

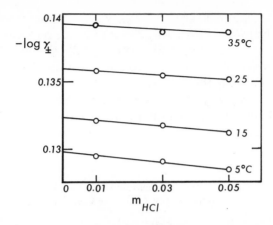

Figure 2. Variation of the activity coefficient of hydrochloric acid with molality in seawater I at a constant ionic strength of 0.66

a sodium chloride solution of the same ionic strength, 0.66 (*19*). In other words, replacing all magnesium chloride, sodium perchlorate, potassium chloride, and calcium chloride in the artificial seawater by sodium chloride without changing the ionic strength has little effect on the activity coefficients of the hydrogen and chloride ions. As this is a 33% change in composition, the constancy of $\gamma_{HCl}{}^{tr}$ is rather remarkable. Furthermore, it suggests that adding sodium sulfate in place of sodium perchlorate and sodium chloride (a change involving 13% of the total ionic strength) should not alter $\gamma_{HCl}{}^{tr}$ significantly when the ionic strength and salinity remain unchanged.

For some purposes, it may prove convenient to refer emf measurements of cell A in seawater to a hypothetical standard state in this solvent medium, defined so that the activity coefficient becomes unity in seawater rather than in an infinitely dilute aqueous medium. This approach is favored by Dyrssen and Sillén (*13*) and Hansson (*8*). A new standard emf, $E^{\circ *}$, on this basis is easily obtained by the relationship:

$$E^{\circ *} = E^{\circ} - \frac{2RT}{F} \ln {}_m\gamma_{\pm} = E^{\circ} - \frac{2RT}{F} \ln \gamma_{\pm}{}^{tr} \qquad (8)$$

The primary medium effect or transfer activity coefficient ${}_m\gamma_{\pm}$ is the value of $\gamma_{\pm}$ in the new standard state (pure seawater), referred to the aqueous standard state; it is therefore identical with $\gamma_{\pm}{}^{tr}$. Values of log $\gamma_{\pm}{}^{tr}$ in seawater I and $E^{\circ *}$ for cell A are summarized in Table IV.

The values of E° used to calculate $E^{\circ *}$ were determined (*20, 21*) by measurements of cell A containing an aqueous solution of hydrochloric acid (molality 0.01 mol/kg). In this way, $E°$ was 0.23421, 0.22865, 0.22242, and 0.21573 V at 5°, 15°, 25°, and 35°C, respectively.

Figure 2 makes it clear that $-\log \gamma_{\pm}$ falls slowly as hydrochloric acid replaces a part of the sodium chloride in seawater I. The linear variation is in accord with Harned's rule, and the magnitude of the drop is a measure of $-\log \gamma_{HCl}*$ referred to the new standard state. The behavioral similarity between seawater I and 0.66M sodium chloride is illustrated by the fact that the values of $\log \gamma_{HCl}*$ for these two solvents at $m_{HCl} = 0.06$ mole/kg are -0.001 and -0.002, respectively.

Table IV.   Trace Activity Coefficient $\gamma_{\pm}{}^{tr}$ of Hydrochloric Acid in Seawater ($E°{}^*$ for Cell A)

$t/°C$	$-\log \gamma_{\pm}{}^{tr}$	$E°*/V$
5	0.1298	0.24853
15	0.1323	0.24377
25	0.1359	0.23850
35	0.1393	0.23276

Buffer Solutions

It appears that one may safely conclude that the relationship between $\log \gamma_{HCl}*$ and solute ionic strength (I_s) will be nearly the same when, instead of hydrochloric acid, small quantities of buffer substances are added in the same amounts to the seawater solvent. At 25°C, this relationship is:

$$\log \gamma^*_{HCl} = 0.016 \, I_s \qquad (9)$$

at a constant total ionic strength of 0.66. For the buffers studied, I_s is equal to m_{NaAc} or m_{BHCl}.

We now turn to emf measurements of cell A containing equimolal buffer solutions ($m_{acid} = m_{salt} = m$) in artificial seawater of the two compositions given in Table I. The emf data are given in Tables II and III. One can write formally:

$$pm_H = \frac{(E-E°^*)F}{RT \ln 10} + \log m_{Cl} + 2 \log \gamma^*_{HCl} \qquad (10a)$$

$$= \frac{(E-E°^*)F}{RT \ln 10} + \log m_{Cl} + 0.032 \, I_s \qquad (10b)$$

Inasmuch as $\gamma_{HCl}*$ in the buffer solutions is not known exactly, m_H derived from Equation 10b should be regarded as the "apparent" molality of hydrogen ion. The experiments described above, however, strongly support the validity of Equation 10b in the artificial seawater without sulfate and, as indicated already, probably also in the seawater with sulfate when the buffer molality is low. In view of the interaction of H^+ and SO_4^{2-}, it is not feasible at this time to demonstrate experimentally the correctness of this assumption.

Figure 3 is a plot of the pm_H at 25°C calculated by Equation 10b as a function of buffer molality m ($m = m_{acid} = m_{salt}$). Data obtained in the sulfate-free seawater I are connected by dashed lines, and those for seawater II (with sulfate) are joined by a solid line. Two features of these results require further study—the separation of the two curves for tris buffers and the separation and curvature of the lines for bis-tris. It seems unlikely that the presence of sulfate can alter the ion–ion interactions sufficiently to cause departures from Equation 9 of the magnitude (0.02 in pm_H at $m = 0.02$) found with tris buffers. Consequently, specific interactions, possibly complexation with cations, are suspected. In this case, the differences in calculated pm_H would be real. As will be seen presently, measurements of cells without liquid junction suggest that this is the case (*see* Table V).

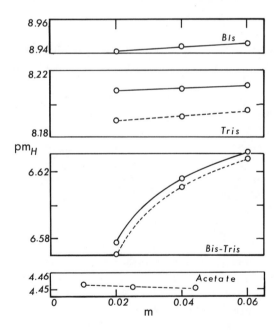

Figure 3. pm_H *(Equation 10b) as a function of buffer molality in artificial seawater at a constant ionic strength of 0.66. (– – –), seawater I; (———), seawater II.*

Table V.   pH and pm_H in Synthetic Seawater at 25°C from
Measurements of Cells Without Liquid Junction (Cell A)
and With Liquid Junction (Cell B)

| | Hydrogen Electrode | | Glass Electrode | |
| | Cell A | Cell B | Cell A | Cell B |
Solution	pm_H	pH	pH	pH
	Seawater I			
HCl, 0.01m	2.000	2.009	1.990	1.993
Acetate, 0.01m	4.454	4.459	4.448	4.456
Tris, 0.02m	8.189	8.181	8.184	8.184
	Seawater II			
Bis-tris, 0.02m	6.577	6.580	(6.573)[a]	6.573
Tris, 0.02m	8.208	8.208	8.206	8.204
Bis, 0.02m	8.942	8.942	8.938	8.938

[a] Reference point

Liquid Junction Potentials

Implicit in the recommendations of Pytkowicz and co-workers for pH measurements in seawater is the belief that the liquid junction potential between seawater of a given salinity and a saturated solution of potassium chloride is independent of the nature and concentration of solutes present at low concentrations in the seawater solvent. This would indeed be the case if seawater is a true constant ionic medium. Hawley and Pytkowicz (5) have estimated that this potential difference amounts to 3.2 mV. As already indicated, this constancy of the liquid junction potential is essential for establishing an experimental scale of pm_H.

To shed further light on the magnitude of the residual liquid junction potential, measurements of cell B were made with the hydrogen electrode (in the cell of Figure 1) and with the glass electrode and a commercial calomel reference electrode in a temperature-controlled cell at 25°C. To obtain added information, this cell also contained a silver–silver chloride electrode, so that the pH could be calculated, relative to a standard, from the emf of the glass/AgCl;Ag cell without liquid junction, much as recommended by Hansson (8). The National Bureau of Standards phthalate and phosphate standard buffer solutions (1) were used as standards for the measurements with a glass electrode. The results are compared with pm_H from Equation 10b in Table V.

The close agreement between pm_H and pH can only be fortuitous and must be attributed to a cancellation of liquid junction potential and activity coefficient terms, as Equation 4 shows. The changes in pm_H and pH from the most acidic to the most alkaline solution are, however, significant. They are as shown in Table VI. It thus seems clear that seawater at $I = 0.66$ (35‰ salinity) effectively nullifies the residual liquid junction potential.

This conclusion was reinforced in another way. The observed emf for the glass electrode–calomel cell (with liquid junction) in eight solutions with pH 2–9 differed by a constant amount from that for the glass–AgCl;Ag combination (without liquid junction) in the same cell. The range was from 9.0 mV in the acetate buffer to 9.6 mV in the tris buffer; the mean was 9.43 mV, and the standard deviation was 0.19 mV.

Measurement of M_H *(Moles of Hydrogen Ion per Kilogram of Seawater)*

The work described in the foregoing sections is of a preliminary nature. Nevertheless, it offers hope that experimental scales of free hydrogen ion concentration (pc_H or pm_H) in seawater may be feasible. One need not know pm_H or pa_H to derive meaningful equilibrium data, such as acid–base ratios and solubilities, provided that suitable apparent equilibrium constants are chosen (7). In these cases, the unit selected for the acidity scale disappears by cancellation. Nevertheless, the acidity of seawater is a parameter of broader impact. It plays a role, for example, in the kinetics of organic oxidation–reduction reactions and in a variety of quasi-equilibrium processes of a biological nature. The concentration of free hydrogen ions is clearly understood, and its role in these complex interactions is more clearly defined than that of a quantity whose unit purports to involve the concept of a single-ion activity.

The emf data for cell A can be used to establish standard reference values of $(pM_H)_S$ in the buffer solutions studied, where the unit of M_H is moles of free H⁺/kg of seawater. With the use of these standards in a form of Equation 6 (with pM_H written in place of pm_H), experimental values of $(pM_H)_X$ in artificial seawater of about 35‰ salinity (and hopefully also in natural seawater of the same salinity) can be obtained. These standard values are listed in Table VII.

The buffer compositions given in the table are molality (m) units, moles/kg of water. Inasmuch as 1 kg of seawater (35‰ salinity) contains 965 g of water, 0.965 mmol of each buffer substance will be required per kilogram of seawater to prepare the buffer solution desired. For greatest precision, the ionic strength should remain always at 0.66. Thus the increased ionic strength caused by adding the buffer salts should be

Table VI. Changes in pm_H and pH

Liquid Junction	Hydrogen Electrode	Glass Electrode
Without	$\Delta pm_H = 6.942$	$\Delta pH = 6.948$
With	$\Delta pH = 6.933$	$\Delta pH = 6.945$

compensated by omitting an equivalent amount of sodium chloride from the artificial seawater used. The error caused by adding the buffer substances to seawater, allowing the ionic strength to increase to 0.67 or 0.68, has not yet been carefully determined, but preliminary measurements indicate that it is probably negligible for most purposes.

Our value (8.224) for the pM_H of the 0.02m tris buffer in seawater II (34.2‰ salinity) at 25°C is considerably higher than that (8.075) for the 0.005m buffer as given by Hansson (8). The buffer ratio is unity in both cases, in the absence of specific interactions as yet unrevealed, and the activity coefficients γ^* are close to unity. Hence, pM_H should be nearly the same as pK^* for tris · H^+ in seawater, a quantity easily derived from $E^{\circ *}$ (Table IV) and E given in Tables II and III. Our results yield $pK^* = 8.185$ in seawater I and 8.205 in seawater II, both values based on the molality scale. On the other hand, Hansson's value of pH_S appears to be based on $-\log (m_H)_t$, where $(m_H)_t$ includes both free hydrogen ion and that combined with sulfate in the form of HSO_4^-. Consequently, pH_S is expected to be lower than our pM_H by about 0.12 unit (22) as found.

Table VII. Standard Reference Values of pM_H in Seawater (M_H = moles hydrogen ion per kilogram of seawater)

$t/°C$	Acetate, $m = 0.01$ [a]	Bis-tris, $m = 0.02$ [b]	Tris, $m = 0.02$ [b]	Bis, $m = 0.02$ [b]
5	4.509	6.953	8.835	9.598
15	4.485	6.766	8.517	9.265
25	4.470	6.593	8.224	8.958
35	4.462	6.431	7.953	8.673

[a] Seawater I—chlorinity 19.1‰, salinity 35.6‰, $I = 0.66$. $m = m_{HAc} = m_{NaAc}$.
[b] Seawater II—chlorinity 19.1‰, salinity 34.2‰, $I = 0.66$. $m = m_B = m_{BHCl}$.

Further study is needed to indicate the validity of the selection of buffer solutions and the assumptions on which these reference standards of pM_H are based. Until then, the values in Table VII should be considered illustrative of the described method and should only be used with a recognition of their tentative character.

Literature Cited

1. Bates, R. G., "Determination of pH," 2nd ed., chap. 4, John Wiley and Sons, New York, 1973.
2. International Union of Pure and Applied Chemistry, "Manual of Symbols and Terminology for Physicochemical Quantities and Units," p. 33, Butterworths, London, 1970.
3. Pytkowicz, R. M., Kester, D. R., Burgener, B. C., Limnol. Oceanogr. (1966) 11, 417.

4. Kester, D. R., Pytkowicz, R. M., *Limnol. Oceanogr.* (1967) **12**, 243.
5. Hawley, J. E., Pytkowicz, R. M., *Mar. Chem.* (1973) **1**, 245.
6. Mehrbach, C., Culberson, C. H., Hawley, J. E., Pytkowicz, R. M., *Limnol. Oceanogr.* (1973) **18**, 897.
7. Pytkowicz, R. M., Ingle, S. E., Mehrbach, C., *Limnol. Oceanogr.* (1974) **19**, 665.
8. Hansson, I., *Deep-Sea Res.* (1973) **20**, 479.
9. Almgren, T., Dyrssen, D., Strandberg, M., *Deep-Sea Res.*, in press.
10. Sørensen, S. P. L., Linderstrøm-Lang, K., *C. R. Trav. Lab. Carlsberg* (1924) **15** (6), article 13.
11. Bates, R. G., "Determination of pH," 2nd ed., p. 447, John Wiley and Sons, New York, 1973.
12. Smith, W. H., Jr., Hood, D. W., "Recent Researches in the Fields of Hydrosphere, Atmosphere, and Nuclear Geochemistry," p. 185, Maruzen Co. Ltd., Tokyo, 1964.
13. Dyrssen, D., Sillén, L. G., *Tellus* (1967) **19**, 113.
14. Lietzke, M. H., Shea, R., Stoughton, R. W., *J. Tenn. Acad. Sci.* (1967) **42**, 123.
15. Pytkowicz, R. M., Kester, D. R., *Oceanogr. Mar. Biol. Ann. Rev.* (1971) **9**, 11.
16. Bates, R. G., Hetzer, H. B., *J. Phys. Chem.* (1961) **65**, 667.
17. Hetzer, H. B., Bates, R. G., *J. Phys. Chem.* (1962) **66**, 308.
18. Paabo, M., Bates, R. G., *J. Phys. Chem.* (1970) **74**, 702.
19. Harned, H. S., Owen, B. B., "The Physical Chemistry of Electrolytic Solutions," 3rd ed., p. 608, Reinhold, New York, 1958.
20. Bates, R. G., Guggenheim, E. A., Harned, H. S., Ives, D. J. G., Janz, G. J., Monk, C. B., Prue, J. E., Robinson, R. A., Stokes, R. H., Wynne-Jones, W. F. K., *J. Chem. Phys.* (1956) **25**, 361.
21. *Ibid.* (1957) **26**, 222.
22. Dyrssen, D., private communication.

RECEIVED October 15, 1974. This work was supported in part by the National Science Foundation under Grant GP-40869X.

11

Analytical Procedures for Transuranic Elements in Seawater and Marine Sediments

HUGH D. LIVINGSTON, DON R. MANN, and VAUGHAN T. BOWEN

Woods Hole Oceanographic Institution, Woods Hole, Mass. 02543

Transuranic elements are extracted from seawater by co-precipitation with either ferric hydroxide or calcium/strontium oxalate or are leached from sediments with 8M nitric acid. Radiochemical separations are used to analyze ^{238}Pu, $^{239,240}Pu$, ^{241}Am, ^{244}Cm, and ^{242}Cm. The electroplated radionuclides are measured by alpha spectrometry using surface barrier detectors. There are spectrometric interferences, especially those arising from natural series radionuclides. Data quality is discussed in terms both of "blank" analyses and of analyses of seawater and sediments containing transuranics which are used in interlaboratory analytical comparisons. Some marine transuranic data support our belief that transuranics sink quite rapidly in the oceans in contrast with "soluble" fallout radionuclides.

Transuranic elements have been introduced to the oceans at various times and rates since the beginning of the nuclear age. At the present time, the major part of the oceanic inventory of transuranics is derived from global fallout of nuclear debris produced in atmospheric testing of nuclear weapons (1). Some local accumulations of these elements are associated with the planned disposal of nuclear waste or releases such as those which follow accidents to aircraft carrying nuclear weapons. The only globally distributed release of a transuranic element from other than weapons testing derived from the malfunction of a satellite carrying a nuclear power source. In April 1964, a navigational satellite (SNAP 9A) burned up in the stratosphere and released ^{238}Pu from its electrical power source mostly into the southern hemisphere. This injection nearly tripled the global inventory of this isotope (1).

Measurement of the concentrations and distributions of transuranic elements in the oceans is becoming increasingly important. At the present time the interest is related to their usefulness as tracers of biogeochemical and geochemical processes. These studies provide knowledge of the eventual fate of these elements in the oceans and the time constants involved in their oceanic pathways. The projected increase in the use of nuclear power towards the latter part of this century necessitates the development of reliable and sensitive analytical techniques of environmental transuranic measurement. This ensures that adequate detection methods are available for environmental hazards resulting from mishandling of nuclear fuels and wastes. This paper outlines the chemical problems encountered in measuring those transuranic elements which have so far been detected in the marine environment.

Only five transuranic elements exist or are anticipated to be produced in amounts which could lead to significant environmental concentrations. These are neptunium (Np), plutonium (Pu), americium (Am), curium (Cm), and californium (Cf). Of these five, only two, plutonium and americium, have been detected and measured already in the marine environment as a result of global fallout of nuclear testing debris. The procedures described below were developed specifically to measure plutonium and americium. However, as will be expanded later, the techniques for measuring americium are also able to detect curium and californium should they be present in significant amounts in the future.

The majority of the longer-lived transuranic nuclides produced by neutron capture reactions decay primarily by α-emission. Most environmental samples contain radionuclides from the natural uranium and thorium series in concentrations often many times greater than transuranic concentrations. As a result, the chemical problems encountered in these measurements are derived from the requirement that separated transuranics should be free of α-emitting natural-series nuclides which would constitute α-spectrometric interferences. Table I lists those transuranic nuclides detected to date in marine environmental samples, together with some relevant nuclear properties. Their relative concentrations (on an activity basis) are indicated although the ratios may be altered by environmental fractionation processes which enrich and deplete the relative concentrations of the various transuranic elements. Alpha spectrometric measurements do not distinguish between ^{239}Pu and ^{240}Pu, so these are reported together. Mass spectrometric measurements of their ratio have been reported for stratospheric samples (2) and for a few marine samples (3). In each case the data indicate that an activity ratio, ^{240}Pu/^{239}Pu, of about 0.8 probably has characterized worldwide integrated fallout. Noshkin and Gatrousis (3) suggest that individual nuclear explosions or test series have been characterized by unique ratios of ^{240}Pu/^{239}Pu and that these might be useful tracers.

Most of the procedures for analysis of transuranic nuclides in seawater and marine sediments have been described in detail elsewhere
both by our laboratory and those of other workers. A full discussion of
these various procedures is found in a comprehensive state-of-the-art
review of techniques proposed for the analyses of transuranic elements
in the marine environment (4). Here we concentrate on the procedures
used at the Woods Hole Oceanographic Institution, the problems encountered, and what is being learned from the data.

Table I.   Transuranic Nuclides Measurable in Marine Environmental Samples

Nuclide	Half-life (yr)	Principal Decay Mode and Energy (MeV)[a]	Typical Activities Relative to $^{239,240}Pu$[b]
^{238}Pu	87.8	α, 5.50, 5.46	3.8
^{239}Pu	24400	α, 5.16, 5.14 ⎫	100[c]
^{240}Pu	6540	α, 5.17, 5.12 ⎬	
^{241}Pu	14.9	β, 0.021	800
^{241}Am	433	α, 5.49, 5.44	20
^{242}Cm	0.22	α, 6.11, 6.07	0.3[d]
^{244}Cm	17.9	α, 5.81, 5.76	0.1[d]

[a] Principal energy first.
[b] These are representative values from our data for marine samples in the Northern
Hemisphere; the relative amounts vary with time, place, and nature of sample, but
these data serve to illustrate general relative concentrations.
[c] Measured together by α-spectrometry.
[d] Only reported from samples contaminated with reprocessed nuclear fuel waste.

Extraction and Concentration

A full account of the problems considered in collecting, storing, and
processing marine samples for transuranic analysis is given in the above-
mentioned review (4). The specific methods discussed here were found
effective at least for the transuranic analyses of seawater and sediments
contaminated by global fallout, nuclear fuel reprocessing wastes, or
nuclear power plant operation waste. In these cases, a preliminary acid
treatment of the sample in the presence of suitable yield monitors seems
to solubilize the transuranic elements and achieves isotopic equilibration
between the yield monitor and sample. The yield monitors used were
either ^{242}Pu or ^{236}Pu for $^{238,239,240,241}Pu$ whereas ^{243}Am was used for
^{241}Am, $^{242,244}Cm$, and by inference, ^{252}Cf. In addition, it was convenient
to use 50 mg of a lanthanide (neodymium) as a carrier for americium to
purify the separated americium fraction.

Seawater. Plutonium and americium analyses were made using 55-l.
seawater samples. The seawater sample is acidified to 0.03M with respect
to hydrochloric acid, and yield monitors and carriers were added. Trans-

uranics were solubilized and equilibrated with the yield monitors in two equally effective ways—bubbling tank nitrogen gas through the acidified sample for several hours at room temperature or stirring the acidified sample by convective heating at about 50°C for several days using aquarium heaters.

The extraction of transuranic elements has been made by co-precipitation in several ways (5, 6). We use either one of two methods, depending on what other nuclides are also sought in the sample. The first method is co-precipitation with 0.5–1.0 g iron as hydroxide at pH 9–10 using ammonium hydroxide while the second method is co-precipitation with calcium and strontium oxalate at pH 5–6 using oxalic acid. There are about 22 g calcium and 0.44 g strontium in 55 l. of open-ocean seawater. Because ^{90}Sr is usually measured in the same seawater sample, we normally add 2 g strontium to that which is naturally present.

The efficiency of these methods of co-precipitation was studied for plutonium, only, by comparing the chemical recovery of the original yield monitor with that of a second (and different) plutonium isotope yield monitor added to the acid solution of the co-precipitated hydroxide or oxalate (6). This second monitor shows all losses following the co-precipitation step. The efficiency of the hydroxide precipitation for plutonium extraction was in the range 70–80%. That for the oxalate co-precipitation was typically 75–85%.

Co-precipitation with Hydroxides. The hydroxides are dissolved in nitric acid and re-precipitated with ammonium hydroxide. A further precipitation is made using ammonia to neutralize most of the acid, then ammonium carbonate to make the solution finally basic. Uranium is retained in the supernate as the carbonate complex.

Co-precipitation with Oxalates. The oxalates are dissolved in nitric acid (1000 ml 16M and 200 ml 24M). At this stage strontium nitrate precipitates and is separated for ^{90}Sr analysis. The solution volume is reduced by evaporation to about 400 ml and then diluted with water to about 1 l. One hundred mg iron is added as carrier, and hydroxides are precipitated with ammonium hydroxide. Two further hydroxide precipitations are made, the second in the presence of carbonate, just as for the hydroxide co-precipitation method.

Sediment. About 50 g of dried sediment was used in transuranic element analyses. Two leachings of the sediment with 200 ml hot 8M nitric acid extracted the plutonium completely and, doubtless, other transuranic elements. Some workers prefer to fuse sediments completely (7, 8), but this is not necessary when the source of transuranic elements is global fallout or nuclear power plant waste. Fusion is probably essential when the transuranic elements in a sample are in relatively resistant oxide particles such as those produced in the destruction by fire of a nuclear device or found in close-in fallout situations.

Purification

The transuranic elements extracted from seawater or sediments are further purified such that:

1. Elements are removed which would electroplate with the transuranic element and cause α-spectral degradation.

2. Natural series radionuclides are removed if the energy of their α-particles interferes with the transuranic elements being measured.

3. Intra-transuranic element separation is made where resolution of the various nuclides being measured is not possible by α-spectrometry alone.

Plutonium Purification. The same purification approach is used for plutonium separated from sediments or seawater. In case reduction may have occurred, the plutonium is oxidized to the quadrivalent state with either hydrogen peroxide or sodium nitrite and adsorbed on an anion exchange resin from $8M$ nitric acid as the nitrate complex. Americium, curium, transcurium elements, and lanthanides pass through this column unadsorbed and are collected for subsequent radiochemical purification. Thorium is also adsorbed on this column and is eluted with $12M$ hydrochloric acid. Plutonium is then eluted from the column with $12M$ hydrochloric acid containing ammonium iodide to reduce plutonium to the nonadsorbed tervalent state. For seawater samples, adequate cleanup from natural-series isotopes is obtained with this single column step so the plutonium fraction is electroplated on a stainless steel plate and stored for α-spectrometry measurement. Further purification, especially from thorium, is usually needed for sediment samples. Two additional column cycles of this type using fresh resin are usually required to reduce the thorium content of the separated plutonium fraction to insignificant levels.

Interferences to Plutonium Measurement. Table II lists plutonium isotopes found in the environment, those used as yield monitors (added in amounts in the range 1–2 disintegrations/min), and the energies of the α-particles produced by their decay, together with the other nuclides which can cause α-spectrometric interference. The interference of ^{234}U to a ^{242}Pu yield monitor is not often serious. If necessary, a correction may be calculated from the ^{238}U α-particles at 4.2 MeV assuming the $^{234}U/^{238}U$ ratio is known. ^{210}Polonium interference from ^{210}Po incompletely removed in plutonium purification or ingrown from ^{210}Pb and/or ^{210}Bi is mostly well enough resolved from $^{239,240}Pu$ or ^{238}Pu. If ^{210}Po is present in amounts such that there is some peak overlap with either $^{239,240}Pu$ or ^{238}Pu, then after dissolution of the plated plutonium, another anion exchange column cycle usually removes this interference completely. ^{228}Thorium remaining in plutonium separated from marine environmental samples (sediments in particular) can seriously interfere with ^{238}Pu measurement. Potential interference is indicated by inspecting the α-spectrum for ^{232}Th, ^{230}Th and/or ingrowth of ^{228}Th daughters, *e.g.*, ^{224}Ra, ^{220}Rn, ^{216}Po, ^{212}Bi, and ^{212}Po. As ^{224}Ra, the daughter of ^{228}Th, is not electroplated, it and its immediate daughter nuclides are not detectable immediately following electroplating. Equilibration of this chain takes two to three weeks, and it is advisable to delay counting for this time if ^{238}Pu is sought. In addition, ^{224}Ra may interfere with any ^{236}Pu used as a yield monitor, and correction for it may be necessary. In either case, correction factors can be calculated from literature branching factors for the decay of the various nuclides in the chain.

Table II. Interferences in Environmental Plutonium Measurements by Chemical Separation and α-Spectrometry

Plutonium Isotope and α Energies (MeV)[a]		Interfering Isotopes and α Energies (MeV)[a]	
^{236}Pu	5.77, 5.72	^{224}Ra	5.68, 5.45[b]
^{238}Pu	5.50, 5.46	^{210}Po	5.31 (directly or by ingrowth)[b]
		^{228}Th	5.42, 5.34
		^{241}Am	5.49, 5.44
$\{$ and ^{239}Pu	5.16, 5.11	^{210}Po	5.31 (directly or by ingrowth)[b]
^{240}Pu	5.17, 5.12		
^{242}Pu	4.90, 4.86	^{234}U	4.77, 4.72

[a] Principal energies first.
[b] Produced by ingrowth from natural series precursors.

Another source of interference in ^{238}Pu measurement can arise from ^{241}Am, which is indistinguishable from ^{238}Pu spectrometrically. Although ^{241}Am originally present in a sample is easily and completely separated from plutonium, its production begins in the separated plutonium through decay of the ^{241}Pu parent. The relative amounts of ^{241}Pu and ^{238}Pu from nuclear fallout at present are such that no serious interference to ^{238}Pu measurement is likely as long as measurement is made within a month or so following separation of plutonium and americium.

^{241}**Plutonium Measurement.** The last-mentioned interference is used indirectly to measure ^{241}Pu. This plutonium isotope is difficult to measure at environmental concentrations by liquid scintillation counting, which is the technique mostly used to measure its low energy β-radiation. Since separated plutonium is freed of americium during chemical purification, ^{241}Am activity on stored plutonium plates increases at a rate controlled by ^{241}Pu. Sufficient ^{241}Am is produced in plutonium separated from relatively plutonium-rich environmental samples to permit its measurement, and by calculation, the ^{241}Pu which produced it, after one to two years delay. It is preferable to separate the newly grown ^{241}Am from plutonium before measurement. Plutonium and americium are removed from a stored plate with 8M nitric acid, and ^{243}Am is added as a yield monitor. After oxidation to the quadrivalent state, plutonium is removed from americium, and iron and/or any thorium is removed from the americium fraction by anion exchange adsorption using two column steps. The first uses an 8M nitric acid medium, the second 12M hydrochloric acid. Both can adsorb plutonium, but the first column removes any thorium still present while the second removes any iron leached from the plate. Americium is electroplated after passage through the chloride column and ^{241}Am measured by α-spectrometry. This approach to ^{241}Pu measurement is described in detail elsewhere (9).

Americium, Curium, and Californium Purification. These elements, together with any lanthanides in the sample or added as carriers, pass through the anion exchange column used to remove plutonium. This fraction is purified to remove natural-series radionuclides which interfere with americium, curium, or californium measurements as well as stable elements which plate with the transuranics and produce spectral degradation. This latter consideration is especially important for lanthanides as neodymium is used as a carrier. Two lanthanide/actinide separation cycles immediately before electroplating are essential for acceptable plate quality.

After plutonium removal, americium, curium, and californium are co-precipitated with neodymium from solution as the oxalate at pH 1. The oxalate is dissolved in nitric acid to decompose oxalate ions, and the hydroxide is precipitated. After acidification and oxidation, the resulting solution is passed through another anion exchange column in the same manner as used earlier for plutonium removal. This column reduces further the concentration of any plutonium and thorium still remaining in the transplutonic fraction after the plutonium removal step. Another anion exchange column, with 1.5M hydrochloric acid, then removes ^{210}Pb, ^{210}Bi, and ^{210}Po. An additional column cycle for plutonium/thorium removal is generally necessary for thorium-rich samples such as sediments. The final steps prior to electroplating are two actinide/lanthanide separation cycles. Americium, curium, and californium are adsorbed on an anion exchange resin from 2M ammonium thiocyanate solution at pH 3–5. Neodymium and any other lanthanides present pass through this column. Americium, curium, and californium are eluted from the column with 4M hydrochloric acid. Following destruction of ammonium thiocyanate traces by nitric acid oxidation, the samples are plated and counted.

Interferences to Americium, Curium, and Californium Measurement. Table III lists the americium, curium, and californium isotopes

present or foreseeable in the environment, together with potential interferences. As [241]Am α-particles are energetically indistinguishable from
[238]Pu α-spectrometrically, [228]Th can also seriously interfere with [241]Am
measurement. The same considerations apply as described above for
[238]Pu with respect to the identification of and correction for this interference. [210]Po from decay of unremoved [210]Bi or ^{210}Pb can interfere with
the [243]Am yield monitor. The 1.5M hydrochloric acid anion exchange
column step seems quite effective in eliminating this interference source.
It is essential for [241]Am measurement that no plutonium be present, as
[238]Pu would interfere with [241]Am. The absence of this interference is
easily established by inspection of the α-spectrum for [239, 240]Pu and [236]Pu
or [242]Pu (whichever yield monitor was used). Also [241]Am may be produced by [241]Pu decay.

**Table III. Interferences in Environmental Am, Cm, and Cf
Measurement by Chemical Separation and α-Spectrometry**

Isotope and α Energies (MeV) [a]	*Interfering Isotope and α Energy (MeV)* [a]
[241]Am 5.49, 5.44	[228]Th 5.42, 5.34
	[238]Pu 5.50, 5.46
	[223]Ra 5.53, several higher energy [b]
[243]Am 5.28, 5.23	[210]Po 5.31 (directly or by ingrowth) [b]
[242]Cm 6.11, 6.07	[212]Bi 6.04, 6.08 [b]
	[227]Th 6.04 (many others) [b]
	[252]Cf 6.12, 6.07
[243]Cm 5.79, 5.74, others ⎱ [244]Cm 5.81, 5.76 ⎰	[224]Ra 5.68, 5.45 [b]
	[227]Th and daughters (complex) [b]
	[236]Pu 5.77, 5.72
[252]Cf 6.12, 6.07	[242]Cm 6.11, 6.07
	[212]Bi 6.04, 6.08 [b]
	[227]Th 6.04 (many others) [b]

[a] Principal energies first.
[b] Produced by ingrowth from natural series precursors.

A further spectral interference with [244]Cm or [252]Cf detection arises
from [227]Ac (from the [235]U natural series). It does not interfere itself,
decaying by emission of weakly energetic β-particles, but [227]Th and a
complex series of daughter nuclides are formed after its decay. They
are effectively removed from the americium fraction by the radiochemical
purification steps used. [227]Actinium remains with the americium fraction
until the actinide/lanthanide separation step, in which it appears to pass
unadsorbed through the thiocyanate column. [227]Th is adsorbed on the
thiocyanate column and elutes with the americium fraction. For this
reason, as little delay as possible should be permitted between the last

thorium removal column operation and the first lanthanide/actinide separation step. For although [227]Th might be thoroughly removed from the americium fraction, it immediately begins to re-equilibrate with its [227]Ac parent. However, as it takes about 18 days to reach one-half of the equilibrium activity value, there is time to separate [227]Ac from the americium fraction, before [227]Th significantly accumulates.

Measurement

Commercially available high resolution surface barrier detectors can be obtained with such low background activity levels for α-detection that measurement of extremely small quantities of α-emitting isotopes is possible. When counting intervals of the order of several days are used, measurement of as little as 1 femtocurie is feasible as our reagent blanks are negligible. Because long counting intervals are required for marine environmental measurements of transuranic nuclides, our capacity for processing an acceptable number of samples is achieved by using a system whereby eight surface barrier detectors (Ortec, Inc.) are used simultaneously. One 1024 channel pulse height analyzer (Northern Scientific NS710) is used for all eight detectors, 128 channels being assigned to each detector. The detector signals are routed to the appropriate set of channels in the analyzer by an eight-input mixer–multiplexer (Northern Scientific NS 459 B).

To preserve the good background characteristics of the detectors (typically about 0.4 count/1000 min in the [239, 240]Pu or [241]Am region), it is necessary to observe certain precautions in their use. Recoil of atoms during α-decay results in the accumulation of recoil atoms within the detector surface (10). When these recoil product atoms are stable or

Table IV.   Transuranic Element

WHOI[b] Values (pCi/kg)

Sample	[238]Pu	[239,240]Pu	[241]Am
Seawater			
SW-I-1	$1.1\times10^{-2}\pm0.3\times10^{-2}$	$7.4\times10^{-2}\pm0.5\times10^{-2}$	—
SW-I-2	$5\times10^{-2}\pm0.3\times10^{-2}$	$26\times10^{-2}\pm0.5\times10^{-2}$	—
Seaweed			
AG-I-1	$4.2\times10^{3}\pm0.3\times10^{3}$	$30\times10^{3}\pm0.2\times10^{3}$	$5.2\times10^{3}\pm0.2\times10^{3}$
Sediment			
SD-B-1[d]	60 ± 5	890 ± 30	231 ± 18

[a] WHOI uncertainty in reported values is the standard deviation of at least six analyses for each intercomparison sample. IAEA uncertainty is the standard error for the values reported by various laboratories after exclusion of outlying values using Chauvenet's criterion.

very long-lived, no serious deterioration of the detector background results. When they are short- or medium-lived, and if they also decay by α-emission, the detector background may be seriously increased. This problem has been encountered frequently with ^{236}Pu measurement. After considerable exposure of a detector to ^{236}Pu, permanent contamination by ^{232}U, ^{228}Th, and successive daughter nuclides takes place. This contamination can severely limit analytical sensitivity for ^{241}Am, ^{238}Pu, ^{242}Cm, ^{244}Cm, or ^{252}Cf. Serious recoil contamination can be prevented by careful control of the nature of the samples exposed to the detector. Yield monitor nuclides should be chosen after considering the recoil contamination risk to detectors. The amount of a nuclide used as a yield monitor should be restricted to the smallest which still gives acceptable chemical yield precision (1–2 disintegrations/min). Furthermore, we check all samples soon after starting measurement in case they contain unexpected amounts of activity which could lead to recoil contamination. Detector backgrounds are monitored at regular intervals to check that the background activity has not increased.

Data and Discussion

The International Atomic Energy Agency organized a series of interlaboratory comparisons for calibration purposes. Those completed so far include two seawater, one seaweed (*Fucus vesiculosus*), and one sediment sample. These materials were contaminated in nuclear waste disposal situations and, in consequence, contain transuranic elements in concentrations much higher than those found in samples contaminated by global fallout of nuclear weapons testing debris. Nevertheless, the data speak directly to questions of calibration of detectors and yield

Interlaboratory Comparisons[a]

IAEA[c] Average Values (pCi/kg)

^{238}Pu	$^{239,240}Pu$	^{241}Am
$0.8\times10^{-2}\pm0.07\times10^{-2}$	$8.7\times10^{-2}\pm0.5\times10^{-2}$	—
$3.5\times10^{-2}\pm0.4\times10^{-2}$	$22\times10^{-2}\pm2\times10^{-2}$	—
$3.8\times10^{3}\pm0.1\times10^{3}$	$27\times10^{3}\pm0.5\times10^{3}$	1) $5.2\times10^{3}\pm0.2\times10^{3}$ [e] 2) $4.4\times10^{3}\pm0.1\times10^{3}$
42 ± 4	960 ± 30	—

[b] Woods Hole Oceanographic Institution.
[c] International Atomic Energy Agency.
[d] Preliminary data.
[e] Average values not reported. Data are those reported by other laboratories.

monitors and in that respect relate to the accuracy of measurements at the lower concentrations encountered in the analysis of marine materials containing "fallout" levels of transuranic elements. Ongoing exercises with this latter type of sample (such as large-volume open ocean seawater) are already producing data which support our belief that accurate data and good interlaboratory analytical agreement are obtainable at those concentrations and at the higher concentrations encountered in waste disposal situations. Table IV presents the results of those interlaboratory comparisons completed to date along with our data in these exercises. The data show that our measurements are of satisfactory quality and that our calibration methods are good.

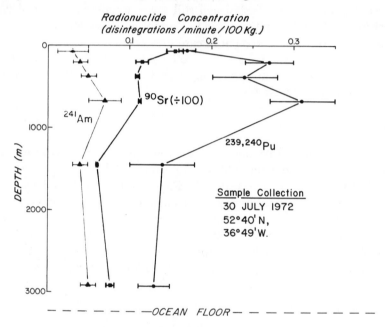

Figure 1. ^{90}Sr, ^{241}Am, and $^{239,240}Pu$ in a North Atlantic water column

Analyses were made on a number of samples of seawater and marine sediments which appear essentially uncontaminated by any artificial radionuclides and which can indicate the total analytical blank found in these measurements. The sediment samples were deep sections of gravity cores from the deep ocean (sections well below the level of the least detectable fallout nuclides), and the water samples were from deep in the Southern Atlantic Ocean. These data may be compared with the analytical data in Figure 1 and Tables V and VI to demonstrate that the analytical blanks are extremely low compared with the nuclide concen-

Table V.   Representative Analysis of Transuranic Elements in a Marine Sediment[a]

Transuranic Nuclide	Sediment Concentration (Disintegration/min/kg(dry wt))[b]
^{238}Pu	3.3 ± 0.6
^{239,240}Pu	52 ± 2
^{241}Pu	430 ± 50
^{241}Am	13.0 ± 0.8
^{244}Cm	n.d.
^{252}Cf	n.d.

[a] Sediment core collected in Buzzards Bay, Mass. 5/24/72; water depth, 16 m; location of sediment section in core, 4–6 cm.
[b] n.d. = not detected.

trations encountered in many areas of the ocean. For seawater, analyses of this type of sample gave concentrations of < 0.005, < 0.005, and 0.007 ± 0.004 disintegrations/min/100 kg for ^{239,240}Pu, ^{238}Pu, and ^{241}Am, respectively. For sediments the corresponding data were 0.2, < 0.1, and 0.04 ± 0.01 disintegrations/min/kg dry sediment, respectively.

^{241}Americium, ^{239,240}Pu, and ^{238}Pu are the only transuranic nuclides measured in seawater where their sole source was global fallout. Figure 1 illustrates the distribution of ^{241}Am and ^{239,240}Pu with water column depth in a set of samples collected in the western North Atlantic Ocean. Also shown are the corresponding ^{90}Sr data. ^{90}Strontium is believed to behave in seawater as a "soluble" fallout radionuclide, moved primarily by physical mixing processes (*11*). It then acts as a tracer of the conservative properties of seawater. ^{90}Strontium concentrations decrease with increasing sampling depth. ^{241}Americium and ^{239,240}Pu concentrations relative to

Table VI.   Distribution of ^{241}Am, ^{239,240}Pu, ^{238}Pu, and ^{241}Pu in a North Atlantic Sediment[a]

Depth of Section from Sediment Surface (cm)	Disintegrations/min/kg (dry sediment)			
	^{238}Pu[b]	$^{239,240}Pu$	^{241}Pu[c]	^{241}Am[c]
0 – 1.3	2 ± 1	160 ± 3	n.m.	43 ± 2
1.3– 2.6	4 ± 1	112 ± 9	1130 ± 160	38 ± 2
2.6– 3.9	2 ± 1	73 ± 6	590 ± 110	26 ± 1
3.9– 5.2	n.d.	47 ± 4	350 ± 70	15 ± 1
5.2– 6.5	n.d.	23 ± 3	n.m.	n.m.
6.5– 7.8	n.d.	15 ± 2	n.m.	n.m.
7.8– 9.1	n.d.	7.8 ± 1.2	n.m.	n.m.
9.1–10.4	n.d.	6.7 ± 0.9	n.m.	n.m.

[a] Sediment collected at water depth of 1115 m with an 8 in. diameter gravity corer on R.V. "CHAIN", Cruise 105, August 6, 1972, at 60°05′ N, 6°02′ W.
[b] n.d. = not detected.
[c] n.m. = not measured.

^{90}Sr increase in deeper samples. This is consistent with our belief that the transuranic nuclides sink more rapidly than ^{90}Sr and in association with sinking particles (*12*).

Table V shows some transuranic radionuclide concentrations found in near shore sediment close to Cape Cod, Mass. The total transuranic content of these shallow sediments agrees well with that predicted as being delivered to the latitude, arguing that the core segment represents part of the period of high ^{238}Pu delivery from SNAP 9A fallout. The implication is that all of the delivered transuranic element is rapidly deposited in the sediment in contrast to the "soluble" fallout radionuclides.

Table VI shows the distribution of ^{238}Pu, ^{241}Pu, ^{239,240}Pu, and ^{241}Am within a sediment core collected several hundred miles northwest of the British Isles. Concentration profiles of plutonium and americium nuclides are rather similar in shape. The transuranic concentrations found in this sediment were surprisingly high. The high concentrations are believed to result from deposition of these nuclides from advected water carrying these nuclides from another area, rather than from the direct vertical transport of sinking particles.

Table VII.   Fraction of Delivered Plutonium Found in Ocean Sediments

Year and Place of Sediment Collection	Depth of Water Overlying Sediment (m)	Fraction of Delivered Plutonium Found in Sediment (%)
1970 41°30' N, 70°50' W	12–24	116 ± 31
1969 41°21' N, 8°41' E	1000	17 ± 5
1971 9°35' S, 12°20' E	1345	32 ± 18
1970 21°54' N, 18°17' W	1410	36 ± 11
1969 39°02' N, 42°36' W	4810	9 ± 3
1971 29°59' S, 4°55' E	4920	6 ± 3
1971 15°49' S, 2°08' E	5349	<2

Our initial series of sediment analyses for transuranic elements led us to believe that there was an inverse relationship between the amounts of transuranic elements present in sediments and the depths of water overlying them (*12*). These data (Table VII) show that progressively smaller fractions of the plutonium estimated to have been delivered to the sea surface are found in sediments in progressively deeper water. More recent data seem to indicate that this relationship is not universally followed. For example, calculations indicated the sediment referred to in Table VI contained ^{239,240}Pu equal to 114% of that predicted as having been delivered to the sea surface at that location. This clearly disagrees with that predicted at that depth from the data in Table VII.

These are examples of the kinds of data that are obtainable for the concentrations of transuranic elements presently in the marine environment. A more complete understanding of the processes which move and remove these elements in the oceans is clearly possible through application of these techniques. It is not necessary to emphasize that this kind of information is and will continue to be necessary in a technological society with an increasing nuclear energy dependence.

Acknowledgment

Many people contributed to this project. They include R. Bojanowski, J. C. Burke, B. L. Dempsey, A. G. Gordon, C. M. Lawson, V. E. Noshkin, J. M. Palmieri, L. D. Surprenant, and K. M. Wong. Their help and efforts are all greatly valued and duly acknowledged.

Literature Cited

1. Hardy, E. P., Krey, P. W., Volchok, H. L., "Global Inventory and Distribution of Fallout Plutonium," *Nature* (1973) **241**, 444.

2. Hardy, E. P., Jr., "Global Atmospheric Plutonium-239 and Plutonium Isotopic Ratios for 1959-1970," *USAEC Health and Safety Laboratory Report*, HASL-273, III-1, 1973.

3. Noshkin, V. E., Jr., Gatrousis, C., "Fallout Plutonium-240 and Plutonium-239 in Atlantic Marine Samples," *Earth Planet. Sci. Lett.* (1974) **22**, 111.

4. "Transuranic Elements," in "Reference Methods for Marine Radioactivity Studies II. Iodine, Ruthenium, Silver, Zirconium, and the Transuranic Elements," *IAEA Tech. Rep. Ser.* (1975) **169**, 5.

5. Hodge, V. F., Hoffman, F. L., Foreman, R. L., Folsom, T. R., "Simple Recovery of Plutonium, Americium, Uranium and Polonium from Large Volumes of Ocean Water," *Anal. Chem.* (1974) **46**, 1334.

6. Livingston, H. D., Mann, D. R., Bowen, V. T., "Double Tracer Studies to Optimize Conditions for the Radiochemical Separation of Plutonium from Large Volume Seawater Samples," in "Reference Methods for Marine Radioactivity Studies II. Ruthenium, Silver, Iodine, Zirconium, and the Transuranic Elements," *IAEA Tech. Rep. Ser.* (1975) **169**, 69.

7. Sill, C. W., "Separation and Radiochemical Determination of Uranium and the Transuranium Elements Using Barium Sulfate," *Health Phys.* (1969) **17**, 89.

8. Aarkrog, A., "Radiochemical Determination of Plutonium in Marine Samples by Ion Exchange and Solvent Extraction," in "Reference Methods for Marine Radioactivity Studies II. Ruthenium, Iodine, Silver, Zirconium, and the Transuranic Elements," *IAEA Tech. Rep. Ser.* (1975) **169**, 91.

9. Livingston, H. D., Schneider, D. L., Bowen, V. T., "^{241}Pu in the Marine Environment by a Radiochemical Procedure," *Earth Planet. Sci. Lett.* (1975) **25**, 361.

10. Sill, C. W., Olson, D. G., "Sources and Prevention of Recoil Contamination of Solid-State Detectors," *Anal. Chem.* (1970) **42**, 1596.

11. Bowen, V. T., Roether, W., "Vertical Distributions of Strontium-90, Cesium-137, and Tritium near 45° North in the Atlantic," *J. Geophys. Res.* (1973) **78**, 6277.

12. Noshkin, V. E., Bowen, V. T., "Concentrations and Distributions of Long-Lived Fallout Radionuclides in Open Ocean Sediments," in "Radioactive Contamination of the Marine Environment," p. 671, IAEA, Vienna, 1973.

RECEIVED January 3, 1975. The U.S.A.E.C. generously supported this work under Contract No. AT(11-1)-3563.

Collection and Analysis of Radionuclides in Seawater

W. B. SILKER

Radiological Sciences Department, Battelle, Pacific Northwest Laboratories, Richland, Wash. 99352

⁷Be and fission products from seawater are collected and their concentrations quantified by analysis. The basic sampling system consists of a unit which filters the water to remove particulate material and then directs the sample flow through a bed of aluminum oxide which retains a determinable amount of the various radionuclides. Fractional radionuclide adsorption was first evaluated in the laboratory under simulated field conditions and then verified by experiments at sea. Analysis of the samples is done by anticoincidence-shielded multidimensional gamma ray spectrometry, which provides a sensitive means of radionuclide detection.

Radioactive material in the world's oceans has provided many useful tracers for investigations of various aspects of oceanography, geochemistry, and other fields of study. These radionuclides include those present in the decay chains of the natural uranium and thorium series, fission products and plutonium isotopes resulting from nuclear weapons testing, and the cosmogenic radionuclides. One of the cosmogenic radionuclides that has proved most useful is ^{7}Be, which is a spallation product resulting from the reaction of cosmic rays with atoms of oxygen and nitrogen.

The techniques used for sample collection and analysis are varied and are dictated by the concentration of the particular radionuclide and the instrumental sensitivity available for its measurement. For those radionuclides that are present in sufficiently high concentration, a 10–100 l. sample of water is collected, and the material is isolated and concentrated by co-precipitation and radiochemical separation. Several *in situ* concentration methods have been developed in which the material of interest is collected on a matrix that is exposed to the water. Folsom (1) uses ferrocyanides of cobalt or copper which are specific for cesium collection to measure the concentration of ^{137}Cs. The amount of water contacted is determined by measuring the quantity of inert cesium, a conservative element, retained in the collector. Jute fiber or sponges impregnated with hydrous ferric oxide were used by Lal (2) to extract silicon and ^{32}Si from tens of tons of seawater. In yet another application, Moore and Reid (3) use acrylic fibers impregnated with manganese oxides to extract radium isotopes from several thousand liters of seawater.

Radionuclide Collection

In our particular case, we are interested in measuring the concentrations of ^{7}Be and the gamma emitting fission products in seawater which are present at levels of a few tenths to hundredths of a dpm/l. To obtain a sufficiently large sample, characteristically 4000 l. of water are processed, and the large-volume water sampler (4) (Figure 1) is used. The bottom section of this unit contains eight parallel filters which remove particulate material greater than 0.3 μ in diameter. Sample flow is then directed through a 0.64-cm thick bed of neutral alumina which retains a determinable fraction of the radionuclides of interest. The alkali metals and alkaline earths which are present in rather high concentrations in the ocean are not adsorbed and pass through the alumina bed. The efficiency with which the aluminum oxide retained various radionuclides was evaluated, both in the laboratory and in the Pacific Ocean. The radionuclides present in the seawater served as tracers for this process. The ocean experiment thus eliminated any inconsistencies resulting from differences in either the chemical or physical state between the prepared solution and the natural states of the various radionuclides.

Laboratory investigations were conducted with a small system that provided prefilters and a 1 $cm^2 \times 0.63$-cm deep bed of aluminum oxide adsorbent. A sample of seawater containing one or more radionuclides of interest was passed through this system at a flow rate of 50 ml/min, which is equivalent to the operational shipboard flow rate of 37.85 l./min for the large ocean sampling system. Collection efficiencies, determined by gamma counting, were evaluated both by comparison of the spike concentrations in the influent and effluent streams and by assay of the amounts retained by the bed with respect to the total throughput. The results from both of these methods were in agreement. The retention efficiencies for the various radionuclides measured by this method are included in Table I. Sorption of the radionuclides was constant at flow rates below 50 ml/min/cm^2 but decreased by 10% at 100 ml/min/cm^2

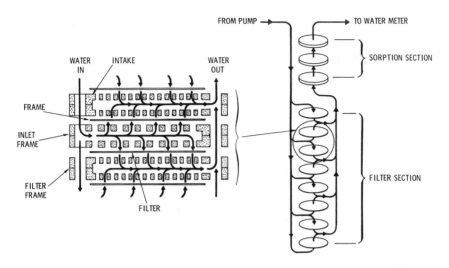

Figure 1.   The Battelle large volume water sampler

and 27% at 150 ml/min/cm². The retention efficiencies were maintained after the equivalent of approximately 6000 l. passed through the ocean system. This is in excess of the 4000 l. normally required to measure adequately the concentration of ocean radionuclides.

Using the facilities aboard the R/V Yaquina, a research vessel operated by Oregon State University, the collection efficiencies of the radionuclides existent in the ocean were determined *in situ*. A large reaction vessel on the ship allowed the trace elements and radionuclides to be separated from about 600 l. of ocean water by precipitation–scavenging reactions while a concurrent water sample was passing through the ocean sampling system. This experiment was repeated at four stations along the Oregon and Washington coasts, which were within the influence of the Columbia River plume yet contained 32% salinity. The Columbia River proximity provided adequate concentrations of radioactive tracers for the experiment.

After the water was collected in the reaction vessel, 288 ppm ferrous sulfate was added. Addition of a stoichiometric quantity (100 ppm) of potassium permanganate thus resulted in formation of hydrous ferric and manganese oxides. Scavenging efficiencies for this precipitation reaction have been reported previously (5, 6). The resulting suspension was agitated for 10–20 min and allowed to settle for 12–24 hr, after which the supernatant liquid was siphoned. The precipitate was removed and

Table I. Percent Uptake of Soluble Radionuclides from Seawater by Alumina[a]

Radionuclide	Laboratory	Ocean Test
^{7}Be	60	66 ± 14
^{46}Sc	43	68 ± 35
^{51}Cr	25	0.5 ± 0.5
^{54}Mn	1.5	44 ± 44
^{60}Co	14	10 ± 6
^{65}Zn	45	44 ± 45
^{95}Zr–Nb	47	59 ± 13
^{106}Ru	18	17 ± 2
^{124}Sb	2.5	3 ± 2
^{144}Ce	58	62 ± 29
^{214}Bi (^{226}Ra)	2.5	4 ± 6
^{234}Th	56	40 ± 7

[a] Those materials that pass on 0.3μ membrane filter are assumed to be soluble. Flow = 50 ml/min/cm^2.

eventually isolated by filtration, dried, and transferred into a suitable counting container. The filters and alumina, which were removed from the ocean sampler, were also dried and transferred into counting containers. All of these samples were analyzed with a multidimensional gamma-ray spectrometer (5), and the amounts of the various radioactive species were determined from the resultant spectra. Bed efficiencies calculated according to the following equation are also included in Table I:

$$\text{Efficiency} = \frac{d/m\ (Al_2O_3)}{d/m\ (\text{ppt}) - d/m\ (\text{filter})} \times 100 \qquad (1)$$

The quantities of radioisotopes used in the laboratory tests were so large that the error caused by counting statistics was less than 3%. Large errors attached to the values from the ocean test arise in part from the counting statistics from the low concentrations of radionuclides involved and in part from the division of a small number by the difference of two relatively large numbers. Good agreement was obtained between the two tests for all isotopes except ^{54}Mn and ^{51}Cr. It is felt that these discrepancies resulted from differences in the chemical or physical form of the isotopes in the two systems, which emphasizes the need for *in situ* evaluation under the conditions from which field samples are expected to be obtained. These comparisons also show that for cases where the natural state of the radionuclides can be duplicated, laboratory evaluation of retention efficiencies are satisfactory.

Currently, we monitor the retention efficiencies in a different manner. Laboratory tests demonstrated that adsorption of the different radioactive species by aluminum oxide obeyed the Freundlich isotherm. Successive beds of the same thickness removed the same fraction of the amount of radionuclide passing the preceding bed as was retained by the upstream bed. In other words, if a bed of a given thickness removed 50% of a particular isotope, a second bed of the same thickness removed 50% of the residual or 25% of the amount initially present. Thus, by using two 0.64-cm beds in series and by measuring the concentrations of radionuclides on each bed, the collection efficiencies for all radionuclides can be calculated easily.

The precision of the sampling method was determined from the analysis of four surface water samples collected at a station in the North Equatorial Atlantic Ocean. Water was simultaneously pumped through two sampling units which were reloaded, and the process was repeated. The analytical results obtained from these four replicate samples, together with the retention efficiencies for the various radionuclides are given in Table II. The range of the replicate measurements agrees reasonably well with that which would be expected from statistical considerations. The poor precisions recorded for ^{144}Ce and ^{226}Ra arise in part from poor counting statistics and also from the fact that only 2.5% of the ^{226}Ra is retained by the aluminum oxide bed.

As previously mentioned, aluminum oxide is used as our primary collector, but other media have been used for special situations. For example, when the Hanford reactors were operating, quantities of hexavalent ^{51}Cr were discharged to the Columbia River and subsequently to the ocean, and we were interested in studying the dispersion rate of the Columbia River plume. Alumina did not collect the dichromate ion efficiently. By using alumina saturated with stannous chloride, the chromium was reduced on contact to the trivalent state, and this was very efficiently retained on the bed. We also found that by saturating alumina with barium sulfate, we could collect radium isotopes, presumably by a replacement reaction with the barium on the matrix. For

Table II.   Measured Radionuclide Concentrations

Sample	^7Be	$^{95}ZrNb$	^{103}Ru	^{106}Ru
1	223 ± 8	68 ± 1	35 ± 3	67 × 5
2	229 ± 8	77 ± 1	51 ± 3	62 ± 6
3	233 ± 7	67 ± 1	39 ± 3	63 ± 8
4	257 ± 8	75 ± 1	42 ± 4	71 ± 5
Average	235.6	72.0	41.8	65.7
Precision (%)	6.2	6.9	16.5	6.5

specialized purposes, we have used beds of other materials—for example, potassium cobalt ferrocyanide for the removal of cesium from seawater. These sampling systems have been used in freshwater where ion exchange resins were utilized for collection and differentiation of cationic, anionic, and non-ionic species existing in the freshwater environment.

Sampling Methods

In our research we want to determine the depth distribution of the various radionuclides in order to determine the vertical eddy diffusivity within the thermocline and the total inventory of the short-lived nuclides within the water column. The equipment used to obtain these samples, pictured in Figure 2, consists of a large motorized reel to lower and retrieve the sampling lines. The sampling lines are a bundle of five

Figure 2.   Water sampling equipment

in Replicate Samples (dpm/m³)

^{144}Ce	^{208}Tl	^{226}Ra	^{234}Th
93 ± 34	3.3 ± 1.5	98 ± 17	7080 ± 200
122 ± 39	3.7 ± 0.4	158 ± 21	7400 ± 230
139 ± 49	3.3 ± 0.6	118 ± 34	6390 ± 200
100 ± 33	3.7 ± 0.5	134 ± 18	7270 ± 230
113	3.5	127	7035
18.6	6.6	20	6.4

2.54-cm diameter polyethylene hoses, each terminating at a different depth, the longest being 100 m. Water is drawn from each depth with deck-mounted centrifugal pumps, and the flow is directed to a sampling unit. The effluent is passed through a standard house water meter to measure the volume of sample processed and is then discharged overboard.

If our particular ship driver is adept at handling his vessel, the hose angle remains at zero, an idealized situation which is generally unattainable. Originally we placed a timed depth recorded at the bottommost sampling location and would determine the other sampling depths by extrapolation. Realizing that this was less than satisfactory and rather than putting moderately expensive equipment at each sampling depth, we developed a very simple system for monitoring depth. Lengths of small bore plastic tubing, terminating at each sampling depth, are pressurized with compressed gas. By measuring the equilibrium pressures, we have a real time monitor of the depths from which samples are actually being drawn.

Sample Analysis

After the samples are returned to the laboratory, they are analyzed nondestructively on highly sensitive anticoincidence-shielded multidimensional gamma-ray spectrometers. An example of one of these counting systems is shown in Figure 3 (7). This particular system has two 11-in. diameter by 6-in. thick sodium iodide for its primary detectors with a 4-in. sodium iodide light pipe, all surrounded by an anticoincidence shield of NE-102 plastic phosphor. The sample to be counted is placed between the two primary detectors. The amplified signals from each detector are fed through separate analog to digital converters for energy and coincidence analysis and then to a 4096 channel memory core which is set up in a 64 × 64-in. matrix array. A single photon interaction in one detector is stored on the corresponding X or Y axis of the memory. Two coincident photons, each reacting with a different detector, are stored in the energy–energy plane at a point uniquely characteristic of their energies. Any event occurring simultaneously within either principal

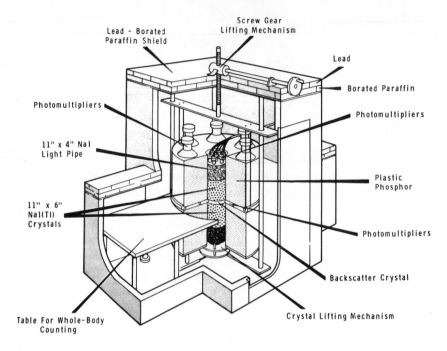

Figure 3. Anticoincidence shielded multidimensional gamma-ray spectrometer

detector and the anticoincidence shield is rejected. These instruments characteristically have a very low background, reduced Compton interference, and good counting efficiency and by coincident gamma-ray analysis provide high sensitivity for the direct measurement of many minor radioactive constituents. The disintegration rate of each radionuclide is calculated from its counting rates by computer solution of a series of simultaneous equations. These equations use Compton correction factors and absolute counting efficiencies predetermined from a standardized source counted in identical geometry. In addition, correction for sample volume, collection efficiency, and decay from time of sampling to analysis are made (8).

The methods of sample collection and analysis that we used are constantly being improved and have proved most satisfactory for our particular requirements. The sampling units are in use by other laboratories across the country in investigations of both fresh and salt water systems.

Literature Cited

1. Folsom, T. R., Sreekuman, C., "Reference Methods for Marine Radioactivity Studies," *IAEA Bull.* (1970).

2. Lal, D., Arnold, J. R., Somayajulu, B. L. K., "A Method for the Extraction of Trace Elements from Seawater," *Geochim. Cosmochim. Acta* (1968) **28**, 1111.
3. Moore, W. S., Reid, D. F., "Extraction of Radium from Natural Waters Using Manganese-Impregnated Acrylic Fibers," *J. Geophys. Res.* (1973) **78**, 8880.
4. Silker, W. B., Perkins, R. W., Rieck, H. G., "A Sampler for Concentrating Radionuclides from Natural Waters," *Ocean Eng.* (1971) **2**, 49.
5. Rosinski, J., Nagamoto, C. T., "Scavenging Radionuclides in Substitute Ocean Water," *Ind. Eng. Chem.* (1960) **52**, 429.
6. Silker, W. B., "Evaluation of the Ferrous-Permanganate System for Water Treatment," *USAEC Rep.* (Jan., 1962), available from Clearinghouse for Federal Scientific and Technical Information, Springfield, Va.
7. Wogman, N. A., Robertson, D. E., Perkins, R. W., "A Large Detector Anti-coincidence Shielded Multidimensional Gamma-ray Spectrometer," *Nucl. Instrum. Methods* (1967) **50**, 1.
8. Silker, W. B., "Horizontal and Vertical Distributions of Radionuclides in the North Pacific Ocean," *J. Geophys. Res.* (1972) **77**, 1061.

RECEIVED January 3, 1975. This work was supported by U. S. Atomic Energy Commission Contract AT(45-1)-1830.

13

Measurement of Organic Carbon in Seawater

PETER J. WANGERSKY

Department of Oceanography, Dalhousie University,
Halifax, Nova Scotia, Canada

*The analysis of seawater for organic carbon is complicated
by the large amount of inorganic carbon present, as well as
by the various complex compounds. All of the methods now
used remove the inorganic carbon by acidification and de-
gassing, thus removing and ignoring the volatile organics.
The standard method involves a wet oxidation with persulfate at elevated temperature and pressure. This method
gives results which disagree with those found by dry combustion techniques. The dry combustion results range between 1.5 and 3.0 times the wet oxidation values and show
considerably greater structure in the organic carbon distribution in the water column. None of the techniques now
available are suitable for routine use.*

The organic carbon present in seawater is usually separated into three
components—the particulate organic carbon (POC), the dissolved
organic carbon (DOC), and the volatile fraction. This division is made
on an operational basis rather than on the basis of chemical or biological
reactivity. Organic material is present in seawater in particle sizes ranging from whales to small organic molecules such as methane; the precise
point at which an organic molecule or collection of molecules stops being
a particle and becomes a solute molecule is not well defined. For analytical purposes, the size at which an aggregation of molecules is considered
to be particulate is defined by the separation method used.

Particulate Organic Carbon

The normal method of separation is filtration. The division between
"dissolved" and "particulate" thus becomes a function of pore size and
filtering characteristics of the particular filter chosen. The pore size

generally considered as the division point is 0.45 μm. This choice is probably a carryover from the days of the first membrane filters. However, this convention, like most such artificial boundaries, is more honored in the breach than in the observance. In order for the small amounts of POC to be measured with any reliability, the filters used must contain very little carbon. The filters generally used are of either glass fiber or silver. The glass fiber filters display a nominal pore size, at best, and trap many particles much smaller than their stated pore size (*1*). The silver filters are much closer to their stated filtration characteristics, but almost no one actually uses the 0.45 μm size. The filters as delivered by the manufacturer contain high and variable amounts of carbon which must be removed by combustion. This combustion alters the pore size of the filters to approximately 0.8 μm. The 0.8 μm filters do not suffer so great a distortion in heating (*2*). As a result, most workers using the silver filters use either 0.8 or 1.2 μm, although 0.45 μm is still considered as the cutoff size for POC. The actual cutoff size is determined by the filter used, and the choice of filters may be determined as much by availability and economics as by filtration characteristics.

There is not too much argument about analysis methods of the collected particulate material. While a few workers continue to use wet oxidation methods, it is acknowledged that these methods are neither as sensitive nor as accurate as the various dry combustion methods. Most of the more recent work has been done with one or another of the commercially available carbon analyzers (*2*). However, a few laboratories still use units assembled before the advent of suitable commercial units (*3*). All of these units, commercial or home-built, oxidize the organic carbon to carbon dioxide at high temperatures and measure this carbon dioxide usually by nondispersive infrared gas analysis (*4*) or by thermal conductivity. The results of the various methods seem completely comparable (*5*).

The only argument on methodology still existing concerns the method of determining the filter blank. Some investigators (*2, 3, 6*) measure the carbon content of filters handled just as the samples, but with no water passed through; other workers (*7, 8*) have attempted to correct for absorption of DOC on the filters by using packs of two or more filters and using the bottom filter, presumably containing only adsorbed organic carbon, as the filter blank. Recent work (*3, 9*) suggests that there is no appreciable adsorption and that the organic carbon measured in the bottom filter of a filter pack is appreciable only in surface waters. Its inclusion in a filter blank results in an underestimation of the POC content of the surface waters. The values normally found for POC range from 0–5 in deep water to 25–100 μg carbon/l. in the surface waters or only a few percent of the total organic carbon present.

Volatile Materials

The volatile fraction is also defined by the measurement technique. As covered in the section on the determination of the dissolved fraction, the methods used for removing carbonate carbon must result in the loss of some fraction of the organic material present. This fraction is customarily considered as the volatile fraction, and it is different if it is lost by acidification and evaporation at 60°C or by acidification followed by freeze-drying. In any case, the volatile fraction is not the fraction volatile at normal surface ocean temperature and pH but rather the fraction volatile at a pH low enough to permit the carbonate removal. In our laboratory, hydrocarbons, ethers, aldehydes, ketones, and some alcohols and acids can be removed by these techniques.

At this time there are no reliable measurements of the volatile fraction as defined by any method. Russian workers (10) have estimated by difference that the fraction removed by evaporation of acidified seawater at 60°C is about 15% of the total DOC. Work has been done on identifying and measuring specific compounds removed from seawater by sparging (11) and by the head space technique (12), but hard numbers and reliable distributions for the whole volatile fraction await the development of a simple, accurate technique.

Dissolved Organic Carbon

The measurement of DOC is a far more difficult task than was originally thought. The difficulties stem in part from the quantities involved; even the most optimistic estimates range only from 1.5 to 3.0 mg carbon/l. The analyses are therefore always in the ppm range. Furthermore, to be useful for any calculations of carbon budgets, the precision and accuracy of the analyses should be better than 5%; 1–2% would be greatly preferred. We must strive for accuracy at the 10 ppb range and be unsatisfied with 50 ppb.

There are methods, generally gas chromatographic or fluorometric, which give us this degree of sensitivity and precision for single compounds or classes of compounds. However, the organic materials in seawater include just about every compound of any biological interest and range in molecular weight from 16 to greater than 100,000. The classes of compounds present range from the simplest of hydrocarbons, methane, to complex polysaccharides containing sulfate and uronic acid residues (13). No single, simple detector or method yet devised is equally sensitive to all of these compounds.

One way to treat this problem of extreme diversity is to convert all compounds to one single compound which can then be measured with sensitivity, accuracy, and precision. The compound normally chosen has been carbon dioxide. It is relatively easy to measure by various techniques, and, as a gas, can be separated from a relatively large sample volume and concentrated to increase sensitivity and precision. However, the carbonate carbon already present in water interferes with so simple an approach.

In fresh water, the amount of carbonate present is low enough so that reasonable accuracy can be achieved simply by measuring the carbonate carbon in one aliquot and the total carbon on another and calculating the organic carbon by difference. In seawater, however, there are over 100 mg of inorganic carbon dioxide for every 7 mg derived from the oxidation of organic carbon. In any determination by difference, the organic carbon present could be less than the analytical error of the method. For this reason, the inorganic carbon is normally removed by lowering the pH well into the acid range, usually with phosphoric acid and then sparging with nitrogen. Even at pH 3, however, carbon dioxide is notoriously reluctant to leave seawater. Since the organic carbon is so small a fraction of the total carbon, care must be taken to ensure that there is no residual carbon dioxide to bias the final measurement. It is in this sparging step that the volatile fraction is lost, although this step defines the volatile fraction.

The small amounts of organic carbon present also increase the difficulties inherent in the sampling process. It is possible, with the proper precautions, to ensure that no organic carbon is added to the samples once they have arrived at the laboratory. It is much more difficult to ensure that the amount of organic carbon in the water in the sample container is the same as the amount originally present in the water mass sampled. Most water samplers are lowered through the water column open and are closed at the sampling depth. To build a sampler which could be lowered closed is possible, but too expensive for routine application. The flushing characteristics of the sampling bottles are good enough so that there is little carry-over of dissolved materials from one depth to the next. However, the highest concentrations of organic matter in the sea are in the surface layer, in the form of hydrocarbons, surface-active compounds, and any material contributed by the research vessel. Running the open bottles through this surface layer results in a considerable carry-over of organic material since these compounds adsorb readily to the polyvinyl chloride used for the construction of sampling bottles (*14*). The amount added to the water samples is not usually significant in measuring total organic carbon, but it may bias measurements of specific compounds.

A greater contamination source is the ship itself. Ocean-going vessels, even research ships, travel in a smog of hydrocarbons and combustion products of hydrocarbons. Besides the omnipresent diesel fuel, the lubricating oils and greases and the various organic preservatives soon coat even the cleanest surface. Clothing worn on cruises commonly retains the odor of diesel fuel through two washings. Fumes from the stacks and the galley often circulate freely through the winch room. Although the shipboard laboratory may be as clean as any shore-based installation, the work done there is limited by what has happened to the sample before it arrives at the laboratory.

If the samples are to be brought back to shore laboratories for analysis, preservation becomes a problem. As a general rule, the addition of preservatives is unsatisfactory. It is difficult to avoid the addition of organic materials along with the preservatives. Acidification, although often advocated, is not a reasonable procedure because some of the organics present can be degraded to volatile fragments by treatment with dilute acids over a long period. Perhaps the only useful storage method is freezing, but this is limited by space requirements, both at sea and in the laboratory. The samples, once thawed, must be analyzed and discarded, but the extra handling involved in thawing and refreezing makes contamination almost a certainty, particularly if many samples are involved.

A problem not originally anticipated was that of preparing proper standards and blanks. Doubly distilled water, even when distilled from permanganate, contains a small amount of organic carbon which may vary with the seasons. A range of 0.5–1.5 mg carbon/l. has been found in singly distilled water in Halifax, with the largest amounts occurring shortly after the spring bloom in the local reservoir system. Halifax city water is derived from acidic peat bogs and is very high in organic matter. Other workers have reported values for distilled water in the range of 0.25 mg carbon/l. (15). In late March, the distilled water in this laboratory contained about 1.2 mg carbon/l. or approximately the level to be expected from surface water in the ocean.

In order to determine whether the method we were using for the determination of dissolved organic carbon was truly linear at very low concentrations, it was necessary to remove the organic carbon at least to the level of sensitivity of our method, about 0.03 mg carbon/l. To do this, we were forced to resort to high temperature combustion of the water in a stream of oxygen (16). Our difficulties in removing the last vestiges of organic matter are convincing evidence that any analysis method which does not indicate a blank of at least 0.25 mg carbon/l. in normally distilled water is suffering from incomplete oxidation.

Some comment should be made concerning the levels of precision and accuracy required of the methods. When we are concerned merely with determining the rough outlines of the oceanic distribution of organic carbon, we can perhaps be satisfied with a precision of ±0.2 mg carbon/l. However, if we wish to make any quantitative calculations of the carbon budget or if we wish to keep track of the amounts of carbon present in different fractions of the organic matter, we must aim for a precision of no worse than ±0.02 mg; ±0.01 mg carbon/l. would be considerably better. If we cannot achieve this degree of precision, some kinds of calculations, particularly of rates of formation and utilization, may simply never be possible.

Wet Oxidation Methods

The analysis methods which have been attempted fall roughly into two categories, wet oxidation and dry combustion. Some attempts have been made to use ultraviolet absorption to measure organic content (17, 18), but these methods are no better than semiquantitative. Of the two classes of oxidation methods, wet oxidation has been the most popular since these methods are not faced with the problems presented by the large amounts of salt in the sample.

The wet oxidation methods all remove the carbon dioxide present in seawater, usually by adding acid and sparging with purified oxygen or nitrogen. A strong oxidant is then added to the solution, and after suitable treatment, usually involving heating, the carbon dioxide generated from the organic carbon is removed from solution and measured. The major variations in method have been in the oxidants used and in the methods of measurement.

There has been a progression to stronger oxidants, with each new oxidant showing a slight increase in the amount of organic carbon present. The oxidants used have included potassium peroxide (19), dichromate in sulfuric acid (20, 21, 22), silver dichromate (23), potassium persulfate at 140°C (4), and hydrogen peroxide and high intensity ultraviolet light (24, 25). The changes in oxidant were made more to speed up the analysis methods than to make the oxidation more complete, but the increase in organic carbon accompanying the shift in methods certainly pointed out the insufficiency of the earlier oxidants. In recent years the persulfate oxidation as described by Menzel and Vacarro (4) has become the standard method in limnology and oceanography.

The methods used for measuring the carbon dioxide generated from the organic carbon have also varied considerably. The DOC values found by wet oxidation methods range from 0.5 to 2.0 mg carbon/l., so that the lower values produce about 22 μg carbon dioxide/ml. The greater the sensitivity of the measuring device, the smaller the amount of sample needed. Early methods included gravimetric and volumetric determinations of the released carbon dioxide. More recently, conductometric (23), gas chromatographic, spectrophotometric, and even mass spectrographic methods have been used. The instrument most commonly used is the nondispersive infrared gas analyzer which has the twin advantages of extreme sensitivity and extreme selectivity (4). The instrument also has the disadvantage of a nonlinear response and thus requires careful calibration. The thermistor detectors used in gas chromatographic adaptations of the method do not suffer from this nonlinearity but are considerably less sensitive. Conceivably, the helium ionization detector could be adapted to this technique; the gain in sensitivity would be considerable, but the cost in time required to ensure the continuity and integrity of the system might outweigh any increased sensitivity.

Although the Menzel and Vacarro (4) technique has been accepted as the standard method in this field and is even incorporated into the chemical oceanographer's manual (26), there are still details of the method which have not been sufficiently tested. In the method as originally published, an aliquot of the sample is pipetted into a combustion vial, the acid and persulfate added, and the mixture bubbled to expel the inorganic carbon. Sharp (27) argued that the addition of persulfate at this point might lead to the oxidation of some of the organic compounds and their expulsion from the solution either as carbon dioxide or as volatile fragments of the originally nonvolatile compounds. He made parallel determinations on a large number of samples from the North Atlantic to measure the effect of the time of persulfate addition; the samples to which the persulfate was added after a preliminary acidification and bubbling with nitrogen to remove carbon dioxide displayed up to 30% more DOC than those run with the normal sequence. Furthermore, the constancy of the deep ocean DOC, so much a feature of the wet oxidation methods, was no longer apparent. The distributions instead seem to show a correlation with water mass distributions (Figure 1).

There are other problems in the wet oxidation method. For one, the precision of the method is far from good. Customarily, three replicates are taken from each water sample. After acidification, bubbling, and the addition of persulfate (not necessarily in that order), the sample ampoules are sealed with a torch. They are then held until it is convenient to heat them, usually for 1 hr at 140°C, break the ampoule, and measure the carbon dioxide evolved. Since there is often a considerable delay between sampling and analysis, sometimes as much as a month, there is no opportunity to retrieve mistakes or accidents. Our own experience has been that wild values, perhaps caused by contamination, occur in about 20% of the samples. The use of either subjective judgment or some empirical rule-of-thumb not far removed from subjective judgment, for the elimination of these wild values is widespread.

However, no matter how precise the method may be made by future refinements, the fact remains that only some fraction of the total DOC is measured. The supposition is often made that ease of chemical oxidation can be equated with ease of biological usage, but this is purely conjecture, easily refuted by examination of a few specific examples. Benzoic acid, for instance, is easily and completely oxidized and yet is used as a bacteriostatic preservative.

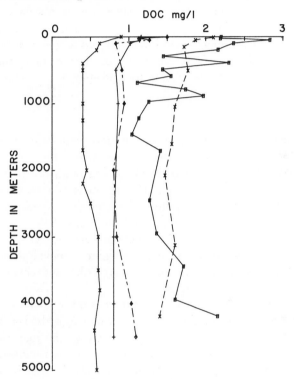

Figure 1.   DOC concentrations in the Sargasso Sea by various methods.   ×—×, wet oxidation (7); +—+, wet oxidation (27); ◇—◇, total combustion (27); □—□, freeze-drying, total combustion (12); × -- ×, evaporation, total combustion (30).

What we can really infer from these data is that the fraction measured by the Menzel and Vacarro (4) variation includes those compounds that are middling difficult to oxidize or volatilize and that the distribution of this fraction is one of high values in the surface waters with low and constant values, within the large variability of the method, throughout the deep oceans. The extra 30% measured by the Sharp (27) variation of the method consists of those compounds degraded to carbon dioxide or to volatile fractions by persulfate at room temperature. No biological correlations can be made with either of these fractions *a priori;* their biological significance must be demonstrated, not assumed.

The whole concept of biological recycling of organic carbon established by Menzel and Ryther (28) rests upon the single foundation of the uniformity of DOC in the deep waters (29, 30, 31). If the true values of DOC and POC vary either with depth or with water mass structure, the rest of their elaborate structure of reasoning collapses. Thus, the validity of the analysis method chosen is more than simply a question of which numbers most closely describe the real world. Our concept of the way in which this piece of the world is put together depends upon the method we choose for our analyses.

Dry Combustion Methods

The obvious alternative to wet oxidation would be some variation of the long familiar dry combustion methods, where the sample is heated to 600°–900°C, usually in a stream of oxygen or in the presence of an oxidizer and the resulting carbon dioxide measured in some manner. While this method has worked admirably through the years for a variety of materials, seawater poses some special problems. Some of these problems have already been discussed. The presence of carbonate in the seawater makes an acidification and carbonate removal step necessary, thus also removing the volatile fraction, and the low concentration of DOC limits the permissible sample size.

The various dry combustion techniques require that the sample be present in the form of dried sea salts. The water must be removed either prior to, or in the course of, the analysis. One of the major differences between methods is the way in which this water is removed. Perhaps the oldest and most straightforward method was devised by Skopintsev and Timofeyeva (10). They simply acidified their samples and dried them at 60°C. The carbonate carbon and the volatiles were removed in the drying process. The sea salts remaining were analyzed for organic carbon by a standard microtechnique. The volatile fraction was defined by this technique as that material which volatilized at 60°C from an acid solution saturated in sea salts. While the advantages of this method are obvious, it is still not completely satisfactory for routine use. The drying step takes too long, limiting the number of samples taken. Also, the possibility of contamination, always present in any organic carbon determination, is increased by the handling necessary in this method.

The DOC values resulting from this technique are startling indeed. Where the wet oxidation methods give surface values between 1 and 2 mg carbon/l., the Russian values range between 1.5 and 3.0 mg carbon/l.

The discrepancies between the methods are even more marked in the deep samples, where the persulfate values are close to 0.5 mg, and the dry combustion values range between 1 and 1.7 mg carbon/l. The discrepancies between the two methods were pointed out almost as soon as enough numbers were in the literature to permit a comparison (*30*). However, North American workers have chosen either to ignore the differences or to assume that the Russian values, although higher, showed the same invariance with depth as was claimed for the American values (*32*). Such an assumption was only possible because the Russian method was tedious, and so few deep samples were taken at any one station that a meaningful depth profile could not be established. Even with the sparse data available, it should have been noted that the difference between the two methods was much greater in the deep water than at the surface.

To settle this controversy, a faster method which offered fewer chances for contamination was obviously needed. In 1967, Van Hall and Stenger (*33*) introduced a direct injection method for DOC. In this method, a sample of solution was injected directly into a combustion furnace in a stream of oxygen, and the resulting carbon dioxide was dried and measured by a nondispersive infrared gas analyzer. This method worked reasonably well for fresh water with a low carbonate and high organic carbon content, well enough so that an instrument quickly became commercially available. The method could not be applied directly to seawater, however, because the lower limit of sensitivity was about the upper limit of seawater DOC. Furthermore, removal of inorganic carbon was necessary for the usual reasons. A later version of the commercial instrument which measured both carbonate carbon and total carbon could not be applied to seawater because the organic carbon, measured as the difference between the two determinations, was less than the combined analytical error of the separate determinations.

For a number of years we attempted to apply variations on this technique to seawater with indifferent success. The major difficulties were the large amounts of water vapor and sea salts involved. The act of injection produced a pressure pulse which had to be damped out, and the amount of dead volume available for damping limited the size of the injection. The small sample size permissible produced an amount of carbon dioxide barely measurable by an infrared analyzer operating at the limit of its sensitivity. With so small a signal, integrated area rather than peak height was the proper measurement. Since the output of the analyzer was nonlinear, it was necessary to linearize the signal before integration. Furthermore, the combination of a pressure pulse on injection and the generally corrosive nature of the sea salt resulted in frequent shattering of combustion tubes, with loss of sample and time. The few bits of data coming out of these attempts seemed to agree with the Russian workers, however.

Jonathan Sharp, at that time a graduate student at Dalhousie, improved this technique until it produced precise, reproducible results. The major changes made in the apparatus were in the shape and size of the combustion tube and in the method of retaining the sea salts. He was able to produce reasonably large peaks with 0.1 ml samples (27). Speed and precision were the advantages of this method. Single injections took about 3 min. Added precision could be obtained by increased replication. Also, the samples could be analyzed almost as soon as they were taken. If an area of the ocean showed interesting anomalies, they would be discovered during the cruise, not two months later.

There are disadvantages to this method, however. It requires that the infrared analyzer, function generator, and integrator operate at the limits of their sensitivity. For an organic carbon content of 2 mg carbon/l., a 0.1-ml sample would contain only 0.2 μg carbon. A precision of 10% would require the instruments to read to the nearest 0.02 μg carbon. The actual precision was somewhat better than 10%, but it was achieved at the cost of large amounts of tender, loving care. This instrument is basically a one-man device and could not, in its present form, become a standard laboratory machine.

A more serious problem is the interpretation of the resulting measurements. We had originally hoped that we might confirm the validity of either the American or the Russian numbers with this method. A comparison of data from the Sargasso Sea, run by the various methods described, shows that the results obtained by this technique fall in between those of the original wet oxidation method and those of the Russian workers (Figure 1). The results are most like those obtained by the Sharp modification of the wet oxidation method, in absolute value if not in detail. Since the combustion temperature used is 900°–1000°C, it seems unlikely that any major portion of the DOC is passing through unscathed. If this method is in fact producing an underestimation of the DOC, it must be the result of escape of organic matter during the acidification and sparging step which removes carbonate carbon. Such an escape might occur by the formation of particulate matter during the bubbling (34). The particles could adhere to the walls of the vessel or might simply be missed in the sampling process.

Another approach to a total combustion is the freeze-drying method of Gordon and Sutcliffe (34). In this method, the seawater sample is acidified with phosphoric acid and taken to dryness in a freeze-drier. The resulting sea salts are analyzed in a commercial C-H-N analyzer. The results obtained by this method are higher than those of Sharp (27) and agree quite well with those of Skopintsev and Timofeyeva (10). Again, the volatile fraction is missing. However, this volatile fraction, again defined by the technique used, is not necessarily the same as that defined by either the Russian workers or by the wet oxidation method.

The necessity for freeze-drying relatively large volumes of water makes this a cumbersome and time-consuming method. It is certainly not a real-time method, any more than is the wet oxidation technique. Also, the dried sea salts are remarkably adsorbent; the use of a perfumed hair dressing or hand lotion by the operator will result in impossibly high values for DOC. In this connection, it is worth noting that the original work on this method was performed with a very old freeze-drier. When it became apparent that the method would work, a new machine was purchased. Samples run on this new machine were hopelessly contaminated. The contamination was eventually traced to vapors produced by the resin used to seal the heating coils to the bottom of the shelf in the freeze-drier.

We have attempted to repeat the work of Gordon and Sutcliffe (*34*) in our own laboratory, with rather mixed results. Replicate samples taken from a station at the edge of the continental shelf give us values which agree closely with those run by Sharp (*27*) using his direct injection method. Even the first batch of samples run by Gordon and Sutcliffe (*34*) may have suffered from some slight contamination and might therefore be too high (*35*).

Shortly after we had run this comparison study, our own aged freeze-drier collapsed into obsolescence. In order to make this method work, the freeze-drier must be specially constructed, without resin in the vacuum chamber and with traps placed in the vacuum line to prevent the back-diffusion of oil vapors from the pump to the vacuum chamber. While we have been awaiting the rejuvenation of our own instrument, rebuilt to these specifications, Michael McKinnon, of our laboratory, has developed a variation of the Russian evaporation method. In this method, as in the freeze-drying method, the great problem is avoiding contamination. Fortunately, when contamination does occur, it seems to affect an entire batch of samples. It is therefore possible to detect the contamination by the judicious use of standards. This method gives values for DOC of the same order as the lowest freeze-drying values or the Sharp (*27*) direct injection values.

Several methods have been developed which seek to escape the inherent nonlinearity of the nondispersive infrared analyzer by the use of other detectors. The flame ionization detector commonly used in gas chromatography has many useful characteristics, including sensitivity and linearity of response, but it does not respond equally to all carbon compounds; it does not respond at all to carbon dioxide. Therefore, the organic compounds must be converted to a single organic compound before measurement. In most of these methods, the organic material is oxidized to carbon dioxide, which is converted to methane and measured with a flame ionization detector (15). While there is no reason why this method should not give results as accurate and precise as any total combustion method, it suffers from the same problems since it requires the removal of carbonates and must use a small sample to cope with the large amounts of water and sea salts.

It appears, then, that the various methods commonly adopted for the measurement of DOC are not necessarily measuring the same quantities. The Menzel and Vacarro method (4) measures those compounds not oxidized by acidic persulfate at room temperature, but oxidized by heating with persulfate for 1 hr at 140°C. The Sharp (27) modification includes the compounds oxidized at room temperature. The Sharp (27) direct injection method presumably measures everything which gets into the combustion chamber, but may lose a small fraction in the bubbling step. The dry combustion methods of Skopintsev and Timofeyeva (10) and Gordon and Sutcliffe (34) measure all of the nonvolatile organic carbon with perhaps some contamination (Figure 1). None of these methods measures the volatile fraction, and this fraction is defined somewhat differently for each method.

The method we choose to adopt affects our entire concept of the cycle of organic carbon in the ocean. The Menzel and Vacarro (4) method presents a picture of a deep ocean of unvarying DOC content. In order to produce such an ocean, the use of organic carbon must be limited to the upper few hundred meters of the ocean, the addition rate of organic carbon to the deep water must be infinitely slow, and the carbon so added must be effectively biologically inert. The Sharp (27) modification of the wet oxidation method and the several dry combustion methods all produce a picture of DOC varying with the water mass distributions. While not enough data have accumulated to allow any real interpretation, it is obvious that the picture of the carbon cycle emerging from these analyses will vastly differ from that produced by the wet oxidation results.

It has been argued that the wet oxidation, although incomplete, still measures the biologically useful DOC. We should keep in mind that this argument was the reason for retaining the permanganate oxidation method after it was discovered that it produced incomplete oxidation. There is a vast store of such data in the literature, all abandoned because it cannot be interpreted. Unless we can discover just which compounds are to be found in the fraction measured by persulfate oxidation and their place in the biological cycle of carbon, we are simply accumulating another such store if we continue to use wet oxidation.

New Directions

We still greatly need a real-time method allowing us to work with reasonably sized liquid samples. We must ultimately be able to analyze for DOC on board ship, so that interesting areas may be resampled and so that time series may be constructed. The plasma spectrograph may be the method we are seeking.

We still do not have a referee method for DOC. The resolution of the disagreement between the various total combustion methods has yet to be accomplished. Probably we will find that the Sharp (27)-direct injection and the Gordon and Sutcliffe (34)-freeze-drying methods are measuring the same quantities once the problem of contamination is solved, and it will be the difference between their values and those of the Russians which has to be resolved.

The measurement of the volatile fraction is very close and will become a routine method within the next few years. If a reasonable total DOC method is developed, the next project will be the qualitative and quantitative analysis of the compounds making up the DOC, and in this work both gas and liquid chromatography will play a large role.

Literature Cited

1. Sheldon, R. W., Sutcliffe, W. H., Jr., *Limnol. Oceanogr.* (1969) **14,** 441.
2. Gordon, D. C., Jr., *Deep-Sea Res.* (1969) **16,** 661.
3. Wangersky, P. J., *Limnol. Oceanogr.* (1974) **19,** 980.
4. Menzel, D. W., Vacarro, R. F., *Limnol. Oceanogr.* (1964) **9,** 138.
5. Gordon, D. C., Jr., *Deep-Sea Res.* (1970) **17,** 233.
6. Sharp, J. H., *Limnol. Oceanogr.* (1973) **18,** 441.
7. Menzel, D. W., *Deep-Sea Res.* (1967) **14,** 229.
8. Banoub, M. W., Williams, P. J. leB., *Deep-Sea Res.* (1972) **19,** 433.
9. Gordon, D. C., Jr., Sutcliffe, W. H., Jr., *Limnol. Oceanogr.* (1974) **19,** 989.
10. Skopintsev, B. A., Timofeyeva, S. N., *Tr. Mar. Hydrophys. Inst.* (1962) **24,** 110.
11. Swinnerton, J. W., Linnenbom, V. J., *J. Gas Chromatogr.* (1967) **5,** 570.
12. Corwin, J. F., in "Symposium on Organic Matter in Natural Waters," *Univ. Alaska Inst. Mar. Sci. Occ. Publ.* (1970) **1,** 169.
13. Hellebust, J. A., "Algal Physiology and Biochemistry," p. 838, W. D. P. Stewart, Ed., Blackwell Scientific Publ., Oxford, 1974.
14. Gordon, D. C., Jr., Keizer, P. D., *Fish. Mar. Service Tech. Rept.* (1974) **481,** 1.
15. Dobbs, R. A., Wise, R. H., Dean, R. B., *Anal. Chem.* (1967) **39,** 1255.
16. Sharp, J. H., Ph.D. Thesis, Dalhousie University, 1972.
17. Ogura, N., Hanya, T., *Nature* (1966) **212,** 758.
18. Mattson, J. S., Smith, C. A., Jones, T. J., Gerchakov, S. M., Epstein, B. D., *Limnol. Oceanogr.* (1974) **19,** 530.
19. Wilson, R. F., *Limnol. Oceanogr.* (1961) **6,** 259.
20. Maciolek, J. A., *U.S. Fish Wildl. Res. Rept.* (1962) **60,** 1.
21. Oppenheimer, C. H., Corcoran, E. F., Van Arman, J., *Limnol. Oceanogr.* (1963) **8,** 487.
22. Krey, J., Szekielda, K. H., *Z. Anal. Chem.* (1965) **207,** 338.

23. Duursma, E. K., *Netherl. J. Sea Res.* (1961) **1**, 1.
24. Beattie, J., Bricker, C., Garvin, D., *Anal. Chem.* (1961) **33**, 1890.
25. Armstrong, F. A. J., Williams, P. M., Strickland, J. D. H., *Nature* (1966) **211**, 481.
26. Strickland, J. D. H., Parsons, T. R., *Fish. Res. Board Can. Bull.* (1972) **167**, 1.
27. Sharp, J. H., *Mar. Chem.* (1973) **1**, 311.
28. Menzel, D. W., Ryther, J. H., in "Symposium on Organic Matter in Natural Waters," *Univ. Alaska Inst. Mar. Sci. Occ. Publ.* (1970) **1**, 31.
29. Riley, G. A., *Adv. Mar. Biol.* (1970) **8**, 1.
30. Wangersky, P. J., *Am. Sci.* (1965) **53**, 358.
31. Craig, H., *J. Geophys. Res.* (1971) **76**, 5078.
32. Menzel, D. W., *Deep-Sea Res.* (1968) **15**, 327.
33. Van Hall, C. E., Stenger, V. A., *Anal. Chem.* (1967) **39**, 503.
34. Gordon, D. C., Jr., Sutcliffe, W. H., Jr., *Mar. Chem.* (1973) **1**, 231.
35. Gordon, D. C., Jr., personal communication, 1974.

RECEIVED January 3, 1975. The research carried out in this laboratory was financed by grants from the National Research Council of Canada. Much of the work was actually done in space furnished by the Atlantic Regional Laboratory of the National Research Council.

C_1–C_3 Hydrocarbons and Chlorophyll a Concentrations in the Equatorial Pacific Ocean

ROBERT A. LAMONTAGNE, WALTER D. SMITH, and
JOHN W. SWINNERTON

U.S. Naval Research Laboratory, Washington, D. C. 20375

*C_1–C_3 hydrocarbons and chlorophyll a analyses were per-
formed on approximately 400 near-surface water samples
(≈ 2 m below the surface) during a one and half month
cruise to the Equatorial Pacific Ocean. The hydrocarbons
were determined by a flame ionization gas chromatograph.
Chlorophyll a was determined fluorometrically. The objec-
tive was to determine if a correlation existed between the
C_1–C_3 hydrocarbons and chlorophyll a in open ocean envi-
ronments. A slight correlation was found between chlorophyll a and ethylene and propylene (R = 0.56). No correlation was found between chlorophyll a and methane, ethane,
or propane.*

The relationship between the biological community and dissolved gases
has been of interest for many years. Much work has been done on
oxygen production and carbon dioxide utilization. Some recent work has
centered around gas production by various kelp communities (1) and
siphonophores (2). However, very little work has been published eluci-
dating the relationship between organisms, primarily one-celled organisms, and dissolved light hydrocarbons (C_1–C_4). Wilson *et al.* (3),
showed the production of carbon monoxide, ethylene, and propylene with
laboratory experiments using dissolved organic carbon produced by
phytoplankton and with an ultradiatom *Chaetoceros galvestonensis*. During a six-month field study of a nearshore environment off Key Biscayne,
Fla., unpublished data by Swinnerton *et al.* (4) showed correlations
between chlorophyll a and carbon monoxide and between light hydrocarbon gas production and chlorophyll a. Zsolnay (5) reported a corre-

lation between chlorophyll a and non-aromatic hydrocarbons for an area off West Africa. During a recent cruise to the equatorial Pacific, chlorophyll a and C_1–C_3 dissolved hydrocarbons were measured to establish whether a correlation existed between dissolved light hydrocarbons and chlorophyll a in the open ocean.

Water samples, taken every 3 hr whenever possible, were obtained from a submersible pump which was mounted approximately 2 m below the surface of the water on the bow of the ship. The method of Swinnerton and Linnenbom (6) was used on all samples for C_1–C_3 hydrocarbon analysis. Chlorophyll a measurements were made following the method for the fluorometric determination of chlorophylls (7).

Results

Figure 1 shows the cruise track from Ecuador to Panama *via* Hawaii and Tahiti onboard the USNS Hayes in March and April 1974. Also shown is a portion of a cruise taken in November and December 1972, onboard the USCGC Glacier, which intersects the 1974 cruise track.

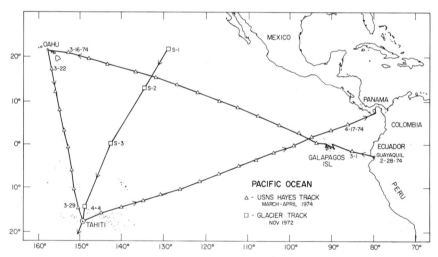

Figure 1.   Cruise track of USNS Hayes, (△), denoting noon position. USCGC Glacier cruise track, (□).

Figure 2 shows the ethylene (C_2H_4) concentrations in the surface waters between Hawaii and Tahiti (Table II). Excluding the values obtained just outside Hawaii (3/21/74) and Tahiti (3/29/74), a broad general increase is apparent with a maximum average value of $\approx 4.0 \times 10^{-6}$ ml/l and a low average value of $\approx 2.4 \times 10^{-6}$ ml/l. Plotted over

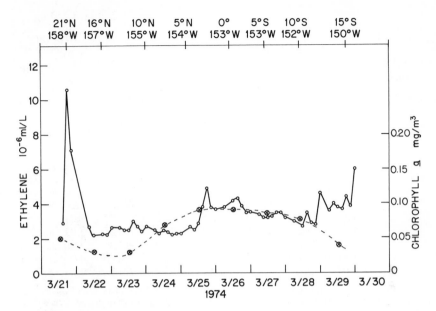

Figure 2.   Surface water ethylene concentrations (○) of USNS Hayes cruise (Hawaii to Tahiti, Table II). Daily averages of chlorophyll a concentrations, (⊗).

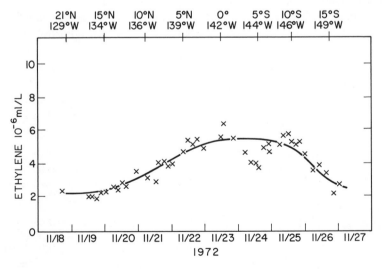

Figure 3.   Surface water ethylene concentrations, (×), of USCGC Glacier cruise between 20°N and 20°S

Table I.   Leg I—

Date	$CH_4{}^b$ (ml/l)	$C_2H_6{}^c$ (ml/l)	$C_2H_4{}^c$ (ml/l)	$C_3H_8{}^c$ (ml/l)
2/28	12.22 ± 9.78	0.51 ± 0.09	11.15 ± 1.90	0.52 ± 0.19
3/1	4.81 ± 0.22	0.51 ± 0.18	8.26 ± 0.75	0.43 ± 0.20
3/2	4.94	0.45	8.02	0.31
3/3	5.32 ± 0.43	0.54 ± 0.07	10.07 ± 1.32	0.63 ± 0.32
3/4	5.44 ± 0.12	0.53 ± 0.02	10.79 ± 1.75	0.39 ± 0.04
3/5	4.71 ± 0.21	0.44 ± 0.12	8.21 ± 1.59	0.30 ± 0.09
3/6	4.40 ± 0.10	0.34 ± 0.07	4.92 ± 0.89	0.22 ± 0.05
3/7	4.36 ± 0.11	0.30 ± 0.08	4.45 ± 0.64	0.20 ± 0.04
3/8	4.45 ± 0.27	0.34 ± 0.08	4.28 ± 1.39	0.24 ± 0.07
3/9	4.33 ± 0.19	0.32 ± 0.19	3.35 ± 0.07	0.17 ± 0.08
3/10	4.45 ± 0.19	0.19 ± 0.03	2.65 ± 0.22	0.14 ± 0.03
3/11	4.30 ± 0.09	0.24 ± 0.07	2.15 ± 0.22	0.10 ± 0.02
3/12	4.12 ± 0.08	0.24 ± 0.08	1.82 ± 0.19	0.07 ± 0.03
3/13	4.09 ± 0.17	0.18 ± 0.04	1.65 ± 0.12	0.04 ± 0.01
3/14	4.11 ± 0.16	0.23 ± 0.07	1.88 ± 0.18	0.05 ± 0.01
3/15	4.11 ± 0.12	0.23 ± 0.04	1.71 ± 0.17	0.06 ± 0.02
3/16	4.03 ± 0.13	0.21 ± 0.05	1.77 ± 0.15	0.04 ± 0.01

[a] Daily averages with one σ standard deviation. Values obtained from $\approx$ 2 m below the surface of the water. In most cases the average is of six or seven values.

these data are the daily average chlorophyll *a* concentrations. The chlorophyll *a* curve exhibits the same general shape as the ethylene curve.

Figure 3 shows the ethylene concentrations reported by Lamontagne *et al.* (7) for the cruise in 1972 which intersects the 1974 cruise track. At that time, an average maximum value for ethylene of $\approx 5.3 \times 10^{-6}$ ml/l, and an average low value of $\approx 2.6 \times 10^{-6}$ ml/l was found. This broad 1972 maximum and the 1974 maximum fit the geographic location of the South Equatorial current relatively well. No chlorophyll *a* data were obtained in 1972.

Tables I, II, and III list the daily averages for Legs I, II, and III with one σ standard deviation for the surface water concentrations of methane

Table II.   Leg II—

Date	$CH_4{}^b$ (ml/l)	$C_2H_6{}^c$ (ml/l)	$C_2H_4{}^c$ (ml/l)	$C_3H_8{}^c$ (ml/l)
3/21	3.93 ± 0.51	0.19 ± 0.18	6.87 ± 3.81	0.04 ± 0.02
3/22	3.90 ± 0.18	0.14 ± 0.03	2.41 ± 0.22	0.09 ± 0.01
3/23	3.92 ± 0.12	0.13 ± 0.03	2.64 ± 0.21	0.05 ± 0.02
3/24	3.89 ± 0.10	0.17 ± 0.05	2.34 ± 0.11	0.09 ± 0.04
3/25	4.12 ± 0.09	0.26 ± 0.05	2.50 ± 0.22	0.08 ± 0.02
3/26	4.02 ± 0.14	0.24 ± 0.04	3.88 ± 0.30	0.08 ± 0.03
3/27	3.88 ± 0.12	0.21 ± 0.05	3.32 ± 0.16	0.05 ± 0.01
3/28	3.85 ± 0.12	0.19 ± 0.05	3.19 ± 0.68	0.04 ± 0.01
3/29	3.81 ± 0.17	0.14 ± 0.05	4.22 ± 0.87	0.05 ± 0.02

[a] Daily average with one σ standard deviation. Values obtained from $\approx$ 2 m below the surface of the water. In most cases the average is of six or seven values.

Ecuador to Hawaii, 1974[a]

C_3H_6[c] (ml/l)	*Chl.* a (mg/m^3)	*Noon Position*
2.15 ± 0.56	0.24 ± 0.08 ⎫	Guayaquil River,
3.11 ± 0.09	0.16 ± 0.04 ⎭	Galapagos Islands
—	—	Marchina Island
2.86 ± 0.33	0.74 ± 0.18	Santa Cruz Island
3.43 ± 0.49	0.12 ± 0.04	0°25'N 93°46'W
2.70 ± 0.34	0.14 ± 0.04	2°40'N 98°24'W
1.65 ± 0.16	0.32 ± 0.17	4°32'N 102°32'W
1.37 ± 0.06	0.12 ± 0.02	6°34'N 107°41'W
1.64 ± 0.53	0.17 ± 0.04	8°29'N 112°30'W
0.99 ± 0.12	0.15 ± 0.06	10°29'N 116°50'W
0.88 ± 0.09	0.13 ± 0.02	12°02'N 122°00'W
0.68 ± 0.11	0.12 ± 0.02	13°48'N 126°56'W
0.59 ± 0.07	0.10 ± 0.03	15°26'N 132°00'W
0.73 ± 0.05	0.06 ± 0.01	16°54'N 136°48'W
0.77 ± 0.11	0.04 ± 0.01	18°31'N 142°41'W
0.73 ± 0.11	0.05 ± 0.01	19°49'N 147°59'W
0.67 ± 0.05	0.05 ± 0.01	20°49'N 153°41'W

[b] 10^{-5} ml/l.
[c] 10^{-6} ml/l.

(CH_4), ethane (C_2H_6), ethylene (C_2H_4), propane (C_3H_8), propylene (C_3H_6), and chlorophyll *a*. The date and noon positions are also listed.

Methane concentrations average $4.22 ± 0.15 × 10^{-5}$ ml/l for the entire cruise. As shown in the tables, methane varies as one changes latitudes and hence crosses currents of varying temperature and salinity. Ethane concentrations under open ocean conditions do not vary greatly as seen in the latter portion of Leg I, where the concentration is $0.26 ± 0.08 × 10^{-6}$ ml/l (away from the Galapagos Islands and the Ecuadorian coast). Leg II exhibits an average concentration of $0.19 ± 0.06 × 10^{-6}$ ml/l. The average concentration for Leg III is $0.25 ± 0.06 × 10^{-6}$ ml/l with values increasing as we approach the Gulf of Panama. The agree-

Hawaii to Tahiti, 1974[a]

C_3H_6[c] (ml/l)	*Chl.* a (mg/m^3)	*Noon Position*
0.52 ± 0.22	0.05 ± 0.01	Pearl Harbor Area
0.88 ± 0.18	0.03 ± 0.01	17°19'N 157°05'W
1.12 ± 0.07	0.03 ± 0.00	12°28'N 150°00'W
0.96 ± 0.15	0.07 ± 0.03	8°05'N 155°00'W
1.14 ± 0.12	0.09 ± 0.02	3°30'N 154°05'W
1.31 ± 0.16	0.09 ± 0.02	0°32'S 153°15'W
1.58 ± 0.07	0.08 ± 0.02	5°46'S 152°39'W
1.40 ± 0.03	0.07 ± 0.01	10°22'S 151°29'W
1.37 ± 0.08	0.04 ± 0.01	14°30'S 150°22'W

[c] 10^{-5} ml/l.
[c] 10^{-6} ml/l.

Table III. Leg III—

Date	$CH_4{}^b$ (ml/l)	$C_2H_6{}^c$ (ml/l)	$C_2H_4{}^c$ (ml/l)	$C_3H_8{}^c$ (ml/l)
4/3	3.75 ± 0.03	0.13 ± 0.02	3.25 ± 0.56	0.06 ± 0.03
4/4	3.76 ± 0.15	0.14 ± 0.04	3.27 ± 0.22	0.06 ± 0.03
4/5	3.85 ± 0.07	0.12 ± 0.03	3.06 ± 0.22	0.05 ± 0.01
4/6	3.84 ± 0.10	0.14 ± 0.03	2.78 ± 0.25	0.05 ± 0.02
4/7	3.80 ± 0.06	0.20 ± 0.09	2.69 ± 0.43	0.10 ± 0.06
4/8	3.70 ± 0.05	0.16 ± 0.03	2.73 ± 0.21	0.06 ± 0.02
4/9	3.79 ± 0.04	0.20 ± 0.06	2.06 ± 0.64	0.07 ± 0.03
4/10	4.05 ± 0.21	0.22 ± 0.05	3.91 ± 0.56	0.09 ± 0.05
4/11	4.04 ± 0.22	0.23 ± 0.06	4.09 ± 0.39	0.09 ± 0.02
4/12	4.23 ± 0.10	0.24 ± 0.08	5.29 ± 0.75	0.13 ± 0.04
4/13	4.42 ± 0.08	0.37 ± 0.06	7.18 ± 0.44	0.23 ± 0.07
4/14	4.65 ± 0.17	0.43 ± 0.05	7.77 ± 0.33	0.23 ± 0.03
4/15	4.64 ± 0.08	0.46 ± 0.10	9.67 ± 1.43	0.26 ± 0.07
4/16	4.57 ± 0.16	0.45 ± 0.04	6.91 ± 1.19	0.18 ± 0.08
4/17	4.47 ± 0.13	0.33 ± 0.08	6.70 ± 1.55	0.15 ± 0.03

[a] Daily averages with one σ standard deviation. Values obtained from $\approx$ 2 m below the surface of the water. In most cases the average is of six or seven values.

ment between Legs I and III in the vicinity of the Galapagos Islands (March 4, 5 and April 14, 15) is good; $0.49 \pm 0.06 \times 10^{-6}$ ml/l and $0.44 \pm 0.07 \times 10^{-6}$ ml/l, respectively.

Ethylene values can vary by almost an order of magnitude as shown in Table I. Concentrations as high as $10.79 \pm 1.90 \times 10^{-6}$ ml/l to a low of $1.71 \pm 0.17 \times 10^{-6}$ ml/l can be found for open ocean conditions. Leg II has been discussed in relation to Figure 2. Leg III exhibits fluctuations between $2.06 \pm 0.64 \times 10^{-6}$ ml/l and $9.67 \pm 1.43 \times 10^{-6}$ ml/l. An over-all average value has not been calculated for ethylene primarily because of these wide variations. Average values of $9.50 \pm 1.64 \times 10^{-6}$ ml/l and $8.72 \pm 0.88 \times 10^{-6}$ ml/l, respectively, were found for the area where Legs I and III intersect.

Propane obtained on Leg I has its greatest value of $0.63 \pm 0.32 \times 10^{-6}$ ml/l in the Galapagos Islands and one of its lowest of $0.04 \pm 0.01 \times 10^{-6}$ ml/l near the Hawaiian Islands. Leg II has an average concentration of $0.06 \pm 0.02 \times 10^{-6}$ ml/l which represents a clean open ocean environment. Concentrations for Leg III vary from $0.06 \pm 0.03 \times 10^{-6}$ ml/l near Tahiti to $0.26 \pm 0.07 \times 10^{-6}$ ml/l in the vicinity of the Galapagos Islands. Comparison of the values between Legs I and III in the area of the Galapagos Islands reveals concentrations of $0.35 \pm 0.06 \times 10^{-6}$ ml/l and $0.25 \pm 0.05 \times 10^{-6}$ ml/l, respectively.

Propylene has the same general characteristics found for the other hydrocarbons. Leg I concentrations vary from $3.43 \pm 0.49 \times 10^{-6}$ ml/l near the Galapagos Islands to $0.59 \pm 0.07 \times 10^{-6}$ ml/l for an area located $\approx$ 1500 miles southeast of Hawaii. Leg II exhibits what appears to be a

Tahiti to Panama, 1974[a]

C_3H_6[c] (ml/l)	Chl. a (mg/m^3)	Noon Position
1.25 ± 0.12	0.03 ± 0.01	Outside Tahiti
1.51 ± 0.19	0.04 ± 0.01	15°21′S 145°12′W
1.49 ± 0.09	0.05 ± 0.01	14°21′S 140°10′W
1.38 ± 0.13	0.05 ± 0.01	12°52′S 135°′47W
1.39 ± 0.18	0.06 ± 0.01	11°35′S 131°04′W
1.12 ± 0.16	0.06 ± 0.01	10°15′S 126°46′W
1.35 ± 0.28	0.06 ± 0.01	8°32′S 121°39′W
1.61 ± 0.17	0.07 ± 0.01	6°43′S 116°18′W
1.61 ± 0.33	0.08 ± 0.01	5°08′S 112°32′W
1.83 ± 0.28	0.10 ± 0.02	3°44′S 108°43′W
2.30 ± 0.17	0.12 ± 0.01	1°55′S 104°14′W
2.24 ± 0.11	0.11 ± 0.04	0°08′S 99°21′W
3.16 ± 0.50	0.09 ± 0.02	1°37′N 95°21′W
1.96 ± 0.32	0.10 ± 0.02	3°25′N 90°30′W
2.06 ± 0.58	0.10 ± 0.02	4°49′N 86°25′W

[b] 10^{-5} ml/l.
[c] 10^{-6} ml/l.

shallow and broad but significant peak. A high concentration of 1.58 ± 0.07 × 10^{-6} ml/l is contrasted with a low value of 0.52 ± 0.22 × 10^{-6} ml/l outside Hawaii and an intermediate value of 1.37 ± 0.08 × 10^{-6} ml/l just outside of Tahiti. Values on Leg III increase steadily from 1.25 ± 0.12 × 10^{-6} ml/l found outside of Tahiti to 3.16 ± 0.50 × 10^{-6} ml/l near the Galapagos Islands. Comparison of the values between Legs I and III near the Galapagos Islands (*see* above for dates) reveals good agreement of 3.06 ± 0.41 × 10^{-6} ml/l and 2.70 ± 0.31 × 10^{-6} ml/l, respectively.

The highest chlorophyll *a* values obtained during the entire cruise was in the area of the Galapagos Islands (0.74 ± 0.18 mg/m³). Values on Leg I decreased to 0.04 ± 0.01 mg/m³ near the Hawaiian Islands. Leg II exhibits a maximum of 0.09 ± 0.02 mg/m³ relative to the average low value of 0.04 ± 0.01 mg/m³ found near Hawaii and Tahiti. Chlorophyll *a* concentrations increased from 0.03 ± 0.01 mg/m³ to 0.12 ± 0.01 mg/m³ as we progressed from Tahiti to the Galapagos Islands. Once again, comparison of values between Legs I and III, where they intersect, shows general agreement; 0.13 ± 0.04 mg/m³ and 0.10 ± 0.03 mg/m³, respectively.

Discussion

Methane data show that this gas in the surface water is in near equilibrium with that in the air. Methane variations do occur in bays, coastal, and anoxic areas, but under open ocean conditions methane concentrations rarely if ever undergo large fluctuations. Methane shows no correlation with chlorophyll *a* or with any of the other light hydrocarbons measured.

Ethane concentrations exhibit the same type of relative constancy as methane. The only variable and high concentrations (relative to open ocean values) of $\approx 0.50 \times 10^{-6}$ ml/l are found near the Galapagos Islands and Ecuador. There is good agreement between the average concentration found in 1972 (0.20×10^{-6} ml/l.) and 1974 (0.28×10^{-6} ml/l.).

Propane is quite similar to ethane in distribution. The highest and most interesting concentrations are also found near the Galapagos Islands and Ecuador. As with ethane, relative agreement is found between the average concentration obtained in 1972 and that found in 1974: 0.30×10^{-6} ml/l and 0.20×10^{-6} ml/l, respectively.

There is a broad but significant increase for ethylene between $\approx 5°$N and $10°$S which corresponds to the position of the South Equatorial Current (S.E.C.). The data reported by Lamontagne et al. (7, Figure 3) shows this maximum for ethylene in the same area. A similar but smaller maximum was found for propylene in 1972, and this is also the case for the 1974 cruise. It is likely that these broad gentle increases for ethylene and propylene result from the S.E.C. sweeping biologically rich upwelled water away from the South American coast. Chlorophyll *a* data follow the general pattern set forth by ethylene and support this speculation.

In regard to the work of Wilson et al. (3) where the only hydrocarbons produced were ethylene and propylene and that of Swinnerton et al. (4), we thought that an analysis of correlation between the unsaturated hydrocarbons and chlorophyll *a* for the entire cruise track would be beneficial. A correlation coefficient, R, of 0.56 was obtained using ethylene and propylene against chlorophyll *a*. A correlation coefficient of 0.59 was obtained between ethylene only and chlorophyll *a*. This latter correlation was done because the ethylene values are much greater than the propylene and also because ethylene was the prominent hydrocarbon gas produced in the work by Wilson et al. (3).

Zsolnay published correlation coefficients for hydrocarbons and chlorophyll *a* of 0.67 (5) and 0.50 (8) for an upwelling area northwest of Africa and for a transect from Nova Scotia to Bermuda, respectively. The chlorophyll *a* concentrations obtained off the coast of Africa in the upwelling region are an order of magnitude greater than ours while those reported for the area between Nova Scotia and Bermuda (0.02–0.81 mg/m^3) are within the same range we obtained for the 1974 cruise. Zsolnay analyzed for nonvolatile hydrocarbons; thus the comparison between his correlation coefficient and ours concerns two different sets of hydrocarbons. In spite of this, the comparison is useful because the values reported by each of us may represent about the maximum correlation coefficient, with R = 1 being the optimum correlation, considering the number of variables encountered. The variability of the size and type of biomass involved, the growth phase of the organisms (*i.e.*, log phase of growth, senescence), season, prevailing current (circulation patterns), and availability of sunlight in the possible photochemical breakdown of organic matter present all contribute to the observed hydrocarbon concentration. An interesting aspect of all this is the fact that comparing all of the hydrocarbons taken on the present cruise to chlorophyll *a* gives a correlation coefficient of 0.50. This indicates that a correlation with chlorophyll *a* may not be applicable under open ocean conditions.

At the present time, we are conducting laboratory experiments to help elucidate whether the major production of these light hydrocarbon gases occurs by the decay of organic matter (dead biomass, *etc.*), photochemical reactions of very labile organic material excreted by organisms, or direct synthesis by the living biomass. This information would be beneficial in assessing whether the light hydrocarbon concentrations are a product of the immediate time and area or whether they are the result of a long-term sequence of events in that given area and/or elsewhere.

A possible area of future work on the correlation between dissolved light hydrocarbons and biomass is the vicinity of the Galapagos Islands. The large and variable concentrations found in that area coupled with the transporting mechanisms there would be ideal for studying production and transport away from source areas.

Acknowledgment

The authors would like to thank J. Haluska and W. Anderson from Old Dominion University, Norfolk, Va. for their collection and analysis of the chlorophyll *a* data. Mechanical assistance from F. Kiselak and G. Bugg from N.R.L. was greatly appreciated.

Literature Cited

1. Loewus, M. W., Delwicke, C. C., *Plant Physiol.* (1963) **38**, 371.
2. Pickwell, G. V., Barham, E. G., Wilton, J. W., *Science* (1964) **140**, 860.
3. Wilson, D. J., Swinnerton, J. W., Lamontagne, R. A., *Science* (1970) **168**, 1577.
4. Swinnerton, J. W., Bunt, J. S., Lamontagne, R. A., unpublished data, 1972.
5. Zsolnay, A., *Deep-Sea Res.* (1973) **20**, 923.
6. Swinnerton, J. W., Linnenbom, V. J., *J. Gas Chromatogr.* (1967) **5**, 510.
7. Strickland, J. D. H., Parsons, T. R., "A Practical Handbook of Seawater Analysis," Fisheries Research Board of Canada, 1972.
8. Lamontagne, R. A., Swinnerton, J. W., Linnenbom, V. J., *Tellus* (1974) **26**, 71.
9. Zsolnay, A., "Hydrocarbon Content and Chlorophyll Correlation in the Waters Between Nova Scotia and the Gulf Stream." Proceedings on Marine Pollution Monitoring—Symposium and Workshop, Gaithersburg, Maryland, 1974.

RECEIVED February 25, 1975

15

Sampling and Analysis of Nonvolatile Hydrocarbons in Ocean Water

R. A. BROWN, J. J. ELLIOTT, J. M. KELLIHER, and T. D. SEARL

Analytical and Information Division, Exxon Research and Engineering Co., Linden, N. J. 07036

Special sampling and analytical methods have been developed to measure dispersed nonvolatile hydrocarbons in open ocean water. Surface samples are collected from moving ships using a stainless steel bucket. Samples are also collected from 3–10 m by drawing water from service lines. For profile samples, specially assembled Niskin bottles are used. In the analytical procedure lipids are extracted from water with carbon tetrachloride. Solvent displacement through a silica gel column then gives a hydrocarbon fraction. Infrared (IR) measurement of this fraction provides a total hydrocarbon value, and then ultraviolet, gas chromatographic, and mass spectrometric analyses give compositional detail. The method is sensitive to 6 μg and reproducible to ±10 relative percent at the 40-μg level.

The role of hydrocarbons in the marine environment is of great interest and the object of considerable study. To know and understand this role is a difficult, complex problem which requires much basic information not now available. For example, we need to know the level and nature of hydrocarbons in open ocean water. This paper describes sampling and measurement methods which were developed and applied to samples collected from tankers and research vessels in the Atlantic and Pacific Oceans during 1972–75. Research vessel sampling was accomplished with the help and cooperation of the Geochemical Ocean Section Study (GEOSECS) project sponsored by the National Science Foundation. These programs were variously founded by Exxon Corp., U.S. Maritime Administration, and the National Oceanic and Atmospheric Administration.

This work began in late 1970 at a time when the concentration of hydrocarbons in ocean water was generally unknown. Initially we attempted to apply the method of Simard *et al.* (*1*) to samples of ocean water. However, this method, which had ample sensitivity for ppm concentrations in refinery effluent water, was not sensitive enough to measure hydrocarbons in ocean water at the ppb (wt) level. Another problem was the presence of interfering organics such as acids and esters. These problems defined the requirements of the analytical method. By now, there are a number of papers to confirm the nature of this problem (*2, 3, 4, 5*).

In developing the method described here, many of the techniques used by others were considered, and some of them were incorporated into the final method. In the overall approach, this method most closely resembles that of Barbier *et al.* (*5*) in which hydrocarbons were measured and partially characterized by compound classes. We should add, however, that the method discussed here includes different overall measurement techniques and is applicable to microgram quantities in contrast to the milligrams available to the above workers.

The method can be used to measure and to characterize hydrocarbons which are classified as nonvolatile (bp > 235°C, $\simeq C_{14}$ and higher molecular weights). It was applied to 3–20-l. unfiltered water samples although the preferred minimum volume is now 8 l.

The method is also useful for measuring extractable organics which include the lipids (esters, acids, hydrocarbons) found in ocean water. This measurement also provides qualitative information about hydrocarbon content. In practice, for example, hydrocarbons frequently occur in the range of 10–40 relative percent of the extractable organics.

Analyses require approximately 0.5 hr for an extractable organic, 3 hr for total hydrocarbons, and several hours for characterization by ultraviolet (UV), gas chromatography (GC), and mass spectrometry (MS).

Sampling

Sampling from a Moving Tanker. Samples were collected from moving tankers and from oceanographic research vessels. When tankers were used, a bucket swung from a boom was used to scoop water from the surface as the tanker traveled at its regular speed of approximately 14 knots. Almost simultaneously a deeper sample at approximately 10 m was taken from the sanitary water line. Every effort was made to prevent contamination. Samples were not taken during ballast or bilge discharge or tank cleaning operations. Extractions were performed on deck or in a room protected from the cargo. When not in use, all equipment was protected from contamination.

An 8-l. stainless steel bucket was used to scoop water from the surface. Bar weights were welded on one side so that the bucket would dip into the water as soon as it touched the surface. The bucket was tied to a previously extracted polypropylene rope and swung from a boom on the windward side of the vessel from as far forward a position as could be safely managed. Upon retrieval of the bucket the water was poured directly into a 1- or 3-gal. glass bottle for immediate extraction by carbon tetrachloride. A carbon tetrachloride rinse of the bucket was part of the extract which was placed in a clean 4-oz bottle with an aluminum-lined cap.

Water from approximately 10 m was obtained by drawing water directly from the sanitary pump line into the extraction bottle. Large volumes of water continuously flow through this line (~ 150 gal/min) during a voyage, and the thoroughly flushed line (20 volume changes per minute) should give a representative sample. One sample was taken every 4–12 hr. In practice, a reasonable variation in concentration is observed.

In addition to precautions taken in sample collection, it is necessary to guard against contamination from the work area while extracting samples aboard ship. This can be accomplished by cleanliness in practice and selecting a clean work area. In favorable weather, for example, extractions are carried out on deck, and at other times, a work room which is free of vapors from either cargo or engine room operations should be used.

During a voyage, it is present practice to collect several blank samples by extracting small water samples under shipboard conditions. Such samples are subjected to the same sources of contamination as actual samples. These blanks have revealed the inadvertent use of impure carbon tetrachloride (identified as rinsing grade). In such cases, the impurity can usually be observed in the IR spectrum routinely used to measure extractable organics. Thus, the spectrum provides a fingerprint to advise the analyst when pure carbon tetrachloride is once again in use. This mixup in use of carbon tetrachloride only occurred in the Atlantic Ocean sampling program.

Sampling from Oceanographic Research Vessel. In sampling from oceanographic vessels, surface samples were taken with a bucket, as was done from tankers. Samples were taken from the windward side after the vessel had left a station and was under way at 2–3 knots.

In the Pacific Ocean, samples were also collected from an uncontaminated seawater line which draws water from approximately 3 m. This line is continuously flushed in normal use and provides a representative sample in the same manner as the sanitary pump line in tankers.

Of 33 stations tested for hydrocarbons to date, a concentration range of
0.0–6.4 ppb was observed. Twenty-two contained < 1 ppb. In addition,
whatever hydrocarbons are detected, they usually appear to be a reasonable portion (0.02–0.24) of the extractable organics. Overall, no measurable contamination occurs in this sample collection procedure.

Profile samples were collected with 30-l. Niskin bottles, using conventional hydrographic sampling techniques, except that no lubrication
of the sampling gear was permitted. Sampling was done by first lowering
the rosette to the deepest water to be sampled. Samples were then taken
stepwise as the rosette was raised toward the surface. This procedure
maximized the flushing of the Niskins. To avoid contamination from the
Niskin bottles, construction specifications were set when the bottles were
ordered. One specification stated that Viton O-rings were to be used in
place of the rubber O-rings normally used. Other specifications eliminated use of a lubricant (silicone oil) and a petroleum resin sealant on
bolt heads.

Before initial use, each Niskin was cleaned with alternate rinses of
isopropanol and distilled water. A given bottle was considered acceptable
only after clean water stored in the bottle and withdrawn for analysis
contained a negligible amount of organics. Thereafter, the Niskins were
cleaned by washing with carbon tetrachloride at the start of a voyage.
During a voyage, the 12 bottles and rosette were used exclusively for
our samples. When not in use, bottles and rosette were covered by a
clean tarpaulin.

Samples collected from aboard oceanographic vessels were poured
into 5-gal. glass jugs for subsequent shipment back to the laboratory for
analysis. Extraction quality carbon tetrachloride was used as a preservative. Considerable care was taken in preparing these jugs for samples.
They were thoroughly cleaned and fitted with previously extracted corks
that were wrapped in freshly cleaned tin foil (aluminum foil is unsatisfactory as it corrodes upon exposure to ocean water). The last carbon
tetrachloride rinse was checked by IR to ascertain that the jug was clean.
After this, jugs were placed in separate wooden crates and shipped to
the research vessel. They may be aboard for several weeks before being
used.

At regular intervals during a voyage, a sample blank was prepared
by pouring only 1–2 l. in contrast to a normal sample of 19 l. into a jug.
This small volume of water covers the carbon tetrachloride to minimize
its loss through evaporation. Each blank was handled as an unknown
sample. To enhance the quality of these blanks, only relatively pure
water collected from deep locations was used.

Table I summarizes sample blanks observed in Atlantic and Pacific
Ocean voyages of the GEOSECS project. In the Atlantic Ocean sampling

program, extractable organic measurements were corrected by subtracting a median value of 2.5 ppb whereas hydrocarbons were corrected downward by 1.0 ppb. During Pacific Ocean sampling some measurements were lower than the median blank values. These values appeared to be more representative of a blank and, therefore, corrections of 1 and 0.5 ppb were applied to all GEOSECS samples. A plausible explanation of why actual blanks may be lower than the measured blanks is that the 5-gal. bottle blanks contain only 1–2 l. of water while sample bottles are completely full. Thus, a blank contains 18 l. of gas space. This would permit breathing of atmospheric air containing hydrocarbons into this gas space. No such gas space is available in actual samples.

The R/V Melville leaked cycloid oil during two different legs of its Pacific Ocean voyage. The cycloid oil was compared with samples collected at those times and was markedly different in composition. Therefore, there was no reason to believe that contamination with leaking cycloid oil had occurred.

Surface samples were collected as usual from aboard the R/V Melville during the GEOSECS cruise for comparison with samples collected from a Zodiac raft which was approximately a quarter of a mile upwind. For a given station, extractable organics and hydrocarbons were very similar as shown in Table II.

Analysis Method

Chemicals. Carbon tetrachloride must have negligible absorption in the 3000 cm⁻¹ region when examined in a 5-cm IR cell. Burdick and Jackson "distilled-in-glass" grade is of satisfactory purity at times, but sometimes it must be purified by passage through a silica gel column. One pint of carbon tetrachloride is passed through a 30×1.5 cm id column containing gel previously dried at 135°C. The first 30 ml is dis-

Table I.   Five-Gal. Glass Jug Sample Blanks

Statistic	Atlantic	Pacific
Number of blanks	10	7
Extractable organics (ppb)[a]		
median	2.5	2.4
mean	2.8	2.8
standard deviation from mean	2.2	1.2
Nonvolatile hydrocarbons (ppb)		
median	1.0	0.9
mean	1.05	0.95
standard deviation from mean	0.4	0.4

[a] Based on 19 l. of water

Table II.   Surface Samples from R/V Melville and Zodiac Raft

Station No.	Position	Sample Collection	Extractable Organics (ppb)	Hydrocarbons (ppb)
223	34°58'N, 151°48'E	R/V	14	2.9
		raft	21	3.3
224	34°15'N, 142°00'E	R/V	12	3.5
		raft	14	2.2
		raft	11	2.2
269	23°58'S, 174°32'W	R/V	15	4.5
		raft	13	2.4

carded, and the remainder is collected. Chloroform, *n*-pentane, and benzene are spectral grade.

To prepare the silica gel (Anasorb, 130–140 mesh, Analabs, Inc.) for use, a portion is placed in an evaporating dish and baked overnight at 275° in a clean oven (15 g, for example, will fill five columns). Longer heating is permissible. When removed from the oven, the gel is immediately poured into a small bottle which is then tightly capped and allowed to cool to room temperature. Activated gel is not allowed to stand more than a day prior to use. Unused gel can be reheated. Hydrocarbon-free nitrogen is tested as part of column blank measurement.

The individual steps of the method are shown in the flow chart of Figure 1. Hydrocarbons are extracted from ocean water and measured by IR spectroscopy. Then the sample is placed on a silica gel column and hydrocarbons are eluted as a fraction(s) which is examined by IR and UV spectrometry, gas chromatography, and mass spectrometry. Each of the principal steps is described in the following sections. Some aspects of this method and typical applications were briefly described by Brown *et al.* (*6*).

Extraction. Simard *et al.* (*1*) previously demonstrated that carbon tetrachloride quantitatively extracts hydrocarbons from refinery waste water. This solvent is advantageous to use because it is transparent to IR in the carbon–hydrogen absorption region of hydrocarbons; thus the extract can be measured directly. To demonstrate the suitability of carbon tetrachloride in this method, numerous blends of hydrocarbons in ocean water at concentrations in the range of 5–20 ppb for 3-l. samples were extracted.

To extract at 3- or 8-l. sample, the sample bottle is rinsed with CCl_4, and this rinse is added to the extract. To extract a 20-l. sample, water is poured into a 1-gal glass jug, and 25 ml CCl_4 is added. Several 1-l. aliquots are extracted, an additional CCl_4 is added to make a total of 125 ml. The sample is vigorously shaken 5 min, and the two phases are allowed to separate. Most of the water is decanted, and the CCl_4 and residual water are poured into a small bottle. The remaining water is then easily removed with a Pasteur pipet. Whenever water is poured from a sample

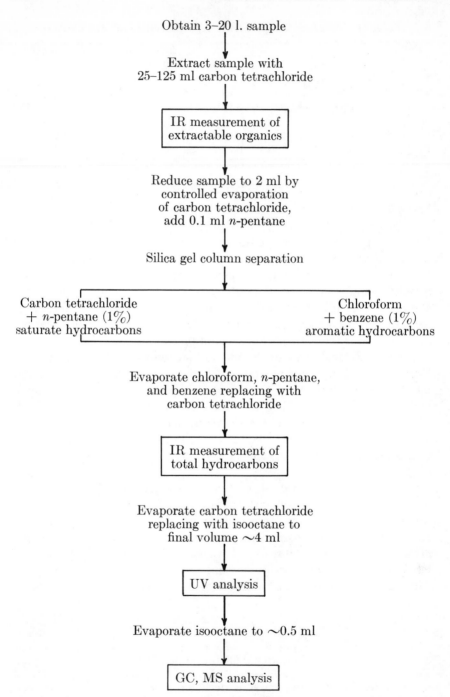

Figure 1.   Analytical method for nonvolatile hydrocarbons in ocean water

bottle into a separate jug for extraction, the emptied sample bottle is rinsed with CCl₄, and the rinse is included as a part of the extract.

Infrared Measurements. An IR measurement is made immediately after extraction and after separation of a hydrocarbon fraction(s) on the silica gel column.

For the first measurement, 5-cm cells with a volume of $\sim$ 15 ml are used. The sample cell is filled with the extract, and the reference cell is filled with CCl₄ from the same batch used for the extraction. The spectrum is scanned from 2600 to 3200 cm⁻¹ using a scale expansion of ×10, if necessary, and a resolution of $\sim$ 3 cm⁻¹. Absorbance of the peak at 2930 cm⁻¹ is then measured using solvent *vs.* solvent as a baseline.

A similar procedure is followed for silica gel column fractions, except that smaller volumes of carbon tetrachloride solution are involved ($\sim$ 0.5 ml), and the IR spectrum is observed using 2-mm microcells.

The IR measurement can be converted to a microgram value by using a calibration factor based on 33 different crude oils. For a 5-cm cell the overall average factor, K_5, was 2.19 l./g cm with a $2\sigma = 28$ relative percent. For measuring hydrocarbons in a microcell, a slightly higher K value of 2.41 l./g cm is used.

Silica Gel Separation. In the procedure as shown in Figure 1, the total carbon tetrachloride extract, which will vary from 20 to 125 ml, is evaporated to 2 ml. This evaporation, carried out in a hood, is carefully conducted on a steam bath with gaseous nitrogen sweeping out vapors.

One-tenth ml *n*-pentane is added, and the sample is poured into the column which is then attached to the manifold (Figure 2). In practice, five of these columns are attached to a manifold—four samples and a

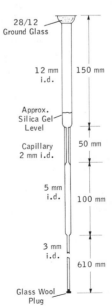

28/12
Ground Glass

12 mm
i.d.

150 mm

Approx.
Silica Gel
Level

Capillary
2 mm i.d.

50 mm

5 mm
i.d.

100 mm

3 mm
i.d.

610 mm

Glass Wool
Plug

Figure 2.   Silica gel column

column blank. Nitrogen pressure of 0.5 lb is applied to allow the sample to be slowly adsorbed into the gel until its surface layer appears to be free of liquid. At this point, the column is disconnected from the manifold, and 5 ml CCl_4 (containing 1 vol. % n-pentane) is added. Once again the column is connected to the manifold, and the CCl_4 is forced into the gel. The column is again disconnected so that 10 ml $CHCl_3$ (containing 1 vol. % benzene) can be added. The column is returned to the manifold, nitrogen pressure is applied, and the liquid is allowed to flow slowly through the gel. Fractions are collected dropwise in small, tared vials. The first 6-ml eluate contains the saturated hydrocarbons. Following this, ~ 4-ml eluate (primarily $CHCl_3$) is collected as the aromatic fraction.

These fractions are combined and carefully evaporated on a steam bath to eliminate the n-pentane, $CHCl_3$, and most of the benzene prior to the IR measurement. A minor amount of benzene does not interfere with the measurements. The ~ 10 ml of solvent is first evaporated to ~ 2 ml and then replenished with 2 ml CCl_4. A second evaporation to 0.5 ml cleans up the solution for a satisfactory IR measurement.

Prior to IR examination the 0.5 ml is weighed to establish its precise volume which is then used along with the IR absorption to calculate micrograms of hydrocarbon.

In practice, four unknown samples and a blank are analyzed as a group. The blank consists of 25–50 ml carbon tetrachloride which is handled as a sample. Small but persistent amounts of hydrocarbons are found for all column blanks and are subtracted from the unknown samples. Blank corrections are also applied to the UV, GC, and MS analyses.

This column blank is not to be confused with sample blanks such as those determined for the 5-gal. jugs. This latter blank correction is applied to (1) extractable organics and nonvolatile hydrocarbon contents measured by the method and (2) UV and GC data. No correction from the sample blank is applied to the compound type analysis by mass spectrometer. Very little net error can be attributed to this omission of a blank correction.

An alternative method is to analyze separately for saturated and aromatic fractions. While this provides somewhat more accurate compositional information, the added complexity is undesirable.

Ultraviolet, Gas Chromatographic, and Mass Spectrometric Measurements. After the silica gel fraction is examined by IR, the carbon tetrachloride solution is changed to isooctane by once again replacing a lower boiling solvent (carbon tetrachloride) with a higher boiling one (isooctane). The UV spectrum is observed over a range of 220–400 nm, using a 1-cm fused silica cell. As used in this method, the UV spectrum only provides qualitative information about the approximate level of dicyclic and higher aromatic hydrocarbons. Zero UV absorption is a reliable indication for absence of aromatics.

Table III.   Gas Chromatographic Measurement of *n*-Paraffins

Sample injection:	50 μl with injector at 340°C
Column:	450 cm long, 0.32 cm od, 2% SE-30 on Chromosorb G
Helium flow:	50 ml/min
Temperature program:	start at 60°C and program at 8°C/min to 275°C
Detector:	flame ionization

Conditions for the GC analysis are shown in Table III. In marine research studies it is desirable to identify *n*-paraffins and the common isoprenoids, pristane, and phytane. For 1-mg samples, this GC column provides high quality chromatograms of well-defined *n*-paraffin peaks with pristane and phytane occurring as easily recognized shoulders to *n*-C_{17} and *n*-C_{18}, respectively. In typical ocean water samples, however, whereas *n*-paraffins are reliably resolved, the isoprenoids may or may not be.

A mass spectrometer is used which is on-line with an IBM 1800 computer. This capability greatly simplifies the extensive calculations which are required. Prior to analysis of a sample(s), a silica gel column blank is analyzed to supply the computer with a spectrum which will subsequently be subtracted from each unknown sample. A blank corrects each set of four unknowns that were simultaneously run through the silica gel separation. In practice, column blanks are quite similar to one another, so a given blank can be used to correct other samples sets also. In each instance the entire fraction is charged to the spectrometer by introducing the sample with a gallium-sealed frit at room temperature and with the sample inlet system open to the vacuum pumps. After 3 min of pumping to remove most of the isooctane, the vacuum valve of the inlet system is closed, and the frit temperature is increased to 315°C. This vaporizes the hydrocarbons from the frit, and the sample spectrum can be obtained.

Table IV.   Evaluation of the Analysis Method

Item	*Observation*
IR calibration at 2930 cm^{-1}	Average coefficient = 2.19; variation from mean (33 crude oils), 2 σ = 28 rel. %; coefficient is applicable to biogenic source hydrocarbons.
CCl_4 extraction	80–111% recovery for known blends of hydrocarbons in ocean water.
Silica gel separation	90–100% recovery for known blends of hydrocarbons in CCl_4.
Precision	σ = 10 rel. % for ~40 μg of hydrocarbons
Sensitivity	6 μg of hydrocarbon

The computer allows more than one method of analysis to be applied to a sample, so two methods are used. In one (7), several different calibration matrices are available to measure hydrocarbons by class, that is, paraffins and cycloparaffins by number of rings (1–6) and 1-ring aromatics. The calibration matrix used in this work is *iso*-C_{24}. As frequently happens, an unknown sample may have a low aromatic content. For such a sample, results from the above method are accepted as its composition.

Samples containing both saturate and aromatics including an observable quantity of dicyclic and higher aromatics are beyond the scope of this method. For the analysis of this sample type, a method is used in

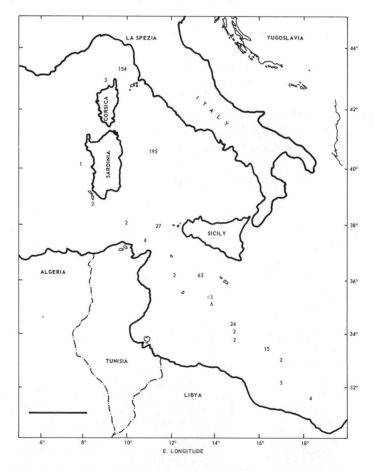

Figure 3.   Nonvolatile hydrocarbons (ppb) in surface water of the Mediterranean Sea

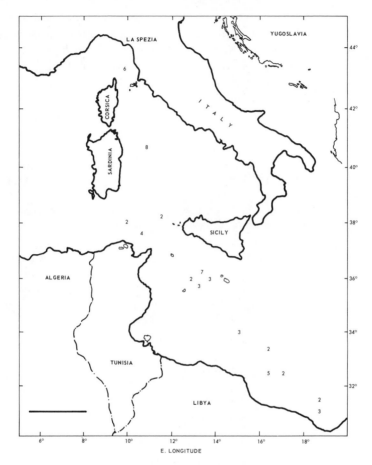

*Figure 4. Nonvolatile hydrocarbons (ppb) in subsurface
(−10 m) water of Mediterranean Sea*

which the mass spectral peak groupings are very similar to those of
Howard *et al.* (8).

Results and Discussion

Considerable work was done to evaluate the efficacy of the method.
Information on the key items of interest is listed in Table IV. The following examples illustrate the nature of results that have been obtained
by the sampling and analytical procedures described.

Nonvolatile Hydrocarbons in the Mediterranean Sea. In August
1972, samples were collected off the Esso Liguria (a LNG ship) during
a ballast and loaded voyage between La Spezia, Italy and Brega, Libya.

Nonvolatile hydrocarbons at the surface (bucket samples) and at 10 m depth are shown in Figures 3 and 4. These data are fairly characteristic of our findings in open ocean water. Surface water shows higher concentrations than that at 10 m. Two of the observed concentrations (195 ppb, 154 ppb) are considerably higher than usually found. Monaghan *et al.* (9) discuss these data in some detail.

Bermuda Deep Sea Profile. Monaghan *et al.* (9) discuss a vertical profile of the water column (surface to bottom, 4500 m) obtained west of the Berumuda Islands using the chartered R/V Panularis. Sampling for extractable organics and nonvolatile hydrocarbons consisted of 13 casts of three 30-l. Niskin bottles. The casts were arranged so that the

Table V.   Extractable Organic and Nonvolatile Hydrocarbons
in Profile Sampling at Bermuda Station[a]

Depth (m)	Extractable Organics (ppb)	Nonvolatile Hydrocarbons (ppb)
0	13, 22	4.5, 4.2
10	15, 8.2	6.3, 3.2
20	10, 12	2.8, 3.1
30	6.3	2.3
50	3.7, 2.4, 3.3	0.5, 0.8, —
75	4.9	1.5
100	4.4, 4.1, 4.2	3.0, 1.0, 1.7
148	3.8	—
195	2.4, 4.3, 2.7	—
298	2.0, 1.7, 4.2	0.9, 0.8, —
400	2.7	—
495	4.8, 3.8, 2.5	1.3, 0.3, —
591	1.8	0.0
620	4.2	—
693	2.2	—
797	6.4	—
899	1.9	0.0
994	2.3, —	0.7, 0.2
1203	2.1	—
1445	2.8	—
2004	2.3	—
2485	2.7	0.1
2527	1.5	0.2
2982[a]	2.9	—
3976[a]	1.4	—
4027[a]	2.1	—
4427[a]	1.9	—

[a] Lat. 32°18′N, Long. 65°32′W, May 14–15, 1972
[b] Extrapolated depth

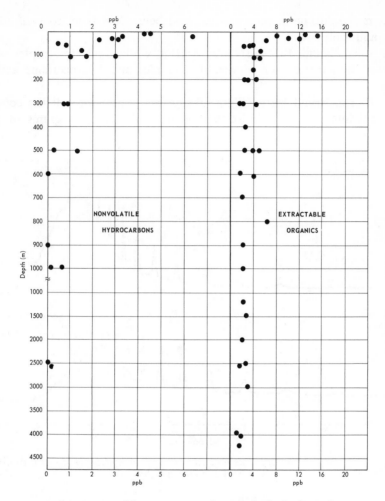

Figure 5. Extractable organics and nonvolatile hydrocarbons at Bermuda Station (32° 18′N, 65° 32′W)

same Niskin bottle was used at those depths where replicate samples were taken. Surface samples were taken with a stainless steel bucket. Carbon tetrachloride was added as preservative when the 5-gal samples were collected. Extraction with carbon tetrachloride was carried out at the Bermuda Biological Station a few days immediately following the sampling.

Extractable organic and hydrocarbon measurements are summarized in Table V and plotted in Figure 5. These data show that concentrations rapidly decline within the first 100 m. During the next several hundred meters the decline is more gradual, and below that point concentrations are uniformly low.

Even though the vessel was drifting while replicate samples were taken, measurement of extractable organics and hydrocarbons agreed reasonably well.

Compositional Data. Figure 6 illustrates the nature of the compositional information provided by the method. Chromatograms of two

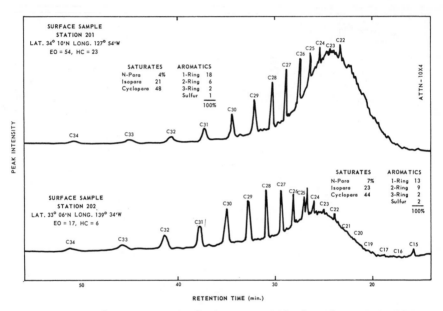

Figure 6. Gas chromatograms and composition of hydrocarbons at GEOSECS Stations 201 and 202

GEOSECS surface samples show a *n*-paraffin distribution with a maximum at C_{28} which is significantly different from the C_{23} maximum observed for the envelope. The hydrocarbons under the envelope are a complex mixture as shown by the hydrocarbon type analysis.

Acknowledgment

Basil Phillips (Exxon International Co.) and D. E. Brandon (Exxon Production Research Co.) assisted in developing the sampling procedures. W. Broecker and John Goddard of the Lamont-Doherty Geological Observatory of Columbia University helped to evaluate profile sampling and other aspects of the work. Laboratory work was carried out by D. E. Bachert and W. D. Henriques.

Literature Cited

1. Simard, R. G., Hasegawa, I., Bandaruk, W., Headington, C. E., *Anal. Chem.* (1951) **23**, 1384.
2. Blumer, M., *Univ. Alaska Inst. Mar. Sci. Occas. Publ.* (1970) **1**.
3. Jeffrey, L. M., *Univ. Alaska Inst. Mar. Sci. Occas. Publ.* (1970) **1**.
4. Levy, E. M., *Water Res.* (1971) **5**, 723.
5. Barbier, M., Joly, D., Saliot, A., Tourres, D., *Deep-Sea Res.* (1973) **20**, 305.
6. Brown, R. A., Elliott, J. J., Searl, T. D., *Natl. Bur. Stand. Spec. Publ.* (1974) **409**, 131.
7. Hood, A., O'Neil, M. J., "Advances in Mass Spectrometry," J. W. Waldron, Ed., p. 175, Pergamon, New York, 1959.
8. Howard, H. E., Ferguson, W. C., Synder, L. R., Thirteenth Annual Conference on Mass Spectrometry and Allied Topics, p. 491, St. Louis, Missouri, May 16-21, 1965.
9. Monaghan, P. H., Brandon, D. E., Brown, R. A., Searl, T. D., Elliott, J. J., Measurement and interpretation of nonvolatile hydrocarbons in the ocean, Part I Measurements in Atlantic, Mediterranean, Gulf of Mexico, and Persian Gulf (1974). Prepared for U.S. Department of Commerce, Maritime Administration Task IV, Contract No. C-1-35049.

RECEIVED January 3, 1975

16

Identification of Specific Organic Compounds in a Highly Anoxic Sediment by Gas Chromatographic–Mass Spectrometry and High Resolution Mass Spectrometry

RONALD A. HITES and WILFRIED G. BIEMANN

Department of Chemical Engineering, Massachusetts Institute of Technology, Cambridge, Mass. 02139

The sediment of the Charles River Basin (an exclusively anaerobic system) was analyzed by computerized gas chromatographic–mass spectrometry (GC–MS) and high resolution mass spectrometry (HRMS). The organic compounds were separated first by methylene chloride extraction followed by gradient chromatography on alumina. This analysis revealed a large number of aliphatic and olefinic hydrocarbons, elemental sulfur, various and abundant polycyclic aromatic hydrocarbons (PCAH) and their alkyl derivatives, and two phthalate esters. Since the PCAH were the most abundant single class of compounds, their identification was pursued in detail. Possible sources of these compounds in the aqueous environment are petroleum, incomplete combustion, anaerobic biosynthesis, or chemical dehydrogenation of natural products.

The Charles River in Boston is not one of the world's most commercially significant rivers. It is, however, of considerable aesthetic importance to metropolitan Boston. Unfortunately, it is highly polluted, a condition which is aggravated by the geometry of the river basin and the associated dams. This has caused the irreversible intrusion of salt water back into the river. Because of its high density, this salt water wedge (shown in Figure 1) prevents the normal seasonal, thermal inversion of the river. As a result very little if any vertical mixing of water occurs. This effect, taken together with the very low flow and with the occasional input of sanitary sewage, has led to a highly anoxic condition of the water. Thus the Charles River may be considered a typical, but extreme case of most bodies of polluted water.

The Charles River is typical of other polluted bodies of water in still another way, namely, there is little detailed information on the specific chemical nature of the pollutants in the water. Previous work in our

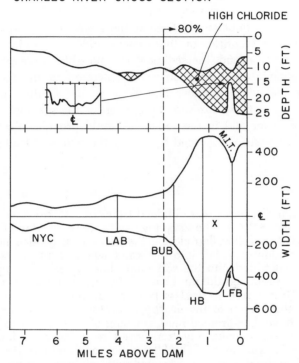

CHARLES RIVER CROSS SECTION

Figure 1. Diagram of the Charles River Basin in Boston. NYC, Newton Yacht Club; LAB, Lars Anderson Bridge; BUB, Boston University Bridge; HB, Harvard Bridge; LFB, Longfellow Bridge. The sediment sampling location is indicated by ×. Data from Ref. 14.

laboratory (*1*) identified some of the specific organic compounds in this water. A number of normal alkanes, several alkyl naphthalenes, a few alkyl anthracenes or phenanthrenes, pyrene, fluoranthene, and phthalate esters were found. The most abundant compounds present were di-(2-ethylhexyl)phthalate and di-*n*-butyl phthalate. Detailed analysis of these data indicated that the alkanes (*see* Figure 2A) were probably of natural origin (from algae and bacteria), the phthalate esters were plasticizers

of ubiquitous urban origin, and the various polycyclic aromatic hydrocarbons were probably of petrochemical origin. The hypothesis was advanced (1) that the aromatic hydrocarbons entered the water by the
urban runoff system and originated in gasoline and other petroleum fuels
used in the urban area.

This paper reports further on the analysis of the Charles River
system by a detailed qualitative examination of the organic compounds
which are present in the river sediment. This investigation of the sediment was undertaken from two points of view. First, the sediment is an
obvious sink for pollutants which enter the river, and it was of interest
to determine the identities of these compounds and, ultimately, their
primary sources. Second, the sediment is a possible pollutant source in
itself, *i.e.*, organic compounds may be generated in the sediment by
chemical or biological reactions and these would, in turn, pollute the
water. In both of these cases, a comparison of the organic compounds
found in the water and in the sediment would provide information on
the natural processes which tend to modify the pollutant load of a body
of water.

Experimental

The sediment sample was obtained from the site indicated in Figure
1 on September 28, 1973. The sampling device was a weighted thin-walled
brass tube with a Teflon flap closing the bottom, suspended on a cotton
rope. It was allowed to fall into the water where it hit the bottom ver-
tically and penetrated into the soft mud about 7 cm. When pulled up
slowly, the flap retained the mud core and some supernatant water. The
sample was transferred to a glass beaker, covered with aluminum foil,
and immediately taken to the nearby laboratory. The sample consisted
of a grey-to-olive, fine grained mud, smelling of hydrogen sulfide. In the
laboratory it was immediately transferred to a large Buchner funnel,
and the water was removed. The filter cake was dried in a desiccator for
two days. The dried mud was extracted with methylene chloride (Nanograde) in a pre-extracted Soxhlet apparatus for 18 hr. The solvent was
evaporated under vacuum at 25°C, and the extract was weighed.

Two batches of the sediment were extracted. The first batch (14.97
g) was broken into pieces 2–10 mm in diameter, and the resulting methylene chloride-extractable matter was 1.92% (0.287 g). The second batch
(19.41 g) was powdered with a mortar and pestle, and the resulting
methylene chloride-extractable matter was 1.96% (0.380 g). There was
no increase in the extractable organic fraction after pulverization, so it
was assumed that the extraction was complete.

The first batch of extract was dissolved in a small volume of methylene chloride and subjected to gradient chromatography on alumina.
Woelm alumina was washed with methylene chloride several times and
then activated at about 95°C for several hours. A 26-ml column (9 mm
id) was prepared with a slurry of this alumina in pentane, and 122 fractions were eluted with a succession of solvents as follows: pentane

(fractions 1–30), 4:1 pentane–benzene (fractions 31–40), 7:3 pentane–benzene (fractions 41–60), 1:1 pentane–benzene (fractions 61–80), benzene (fractions 81–102), 1:1 benzene–methanol (fractions 103–112), and methanol (fractions 113–122). All solvents were nanograde and were shown to be pure enough for this application by GC analysis.

Each fraction was evaporated to dryness in 10-ml pear-shaped flasks, redissolved in a small amount of methylene chloride, and gas chromatographed. GC conditions were as follows: column, 6 ft $\times$ 0.125 in. od stainless steel packed with 3% OV-17 on 100-120 mesh Gas-Chrom Q; temperature program, 70–330°C at 12°/min, holding at the final temperature for up to 20 min; and carrier gas flow rate, 28 ml/min. Fractions containing very little material or with GC patterns very much like adjoining fractions were combined. These fractions were then analyzed on the GC–MS computer system to obtain mass spectral data on the individual components. For certain fractions additional information was gained from ultraviolet spectra.

The second batch of extract was fractionated into pentane-soluble and pentane-insoluble components; 31% of the methylene chloride extract was soluble in pentane. This pentane-soluble fraction was then analyzed by HRMS by directly inserting the sample into the ion source.

The GC–MS computer system consists of a Perkin-Elmer 990 gas chromatograph interfaced to a Hitachi RMU-6L mass spectrometer which is in turn interfaced to an IBM 1802 computer. The details of this hardware and associated software have been published previously (2). The HRMS system consists of a DuPont Instruments 21-110B mass spectrometer and a D. W. Mann comparator interfaced to the IBM 1802 computer. Details on this system are available elsewhere (3). Both mass spectrometers were operated at 70 eV ionizing energy.

Results and Discussion

In general, the compounds which were identified in the sediment of the Charles River Basin can be divided into three groups. The first group comprises fractions 1–30 from the alumina chromatography and contains aliphatic and olefinic hydrocarbons and elemental sulfur. The second group is made up of fractions 31–91 and contains polycyclic aromatic hydrocarbons. The third group consists of fractions 92–122 and contains phthalate esters.

Aliphatic and Olefinic Hydrocarbons. Of the first 30 fractions, only fractions 4–7 contained significant quantities of material. The exploratory gas chromatographic analysis of these four fractions showed them to be very similar; they were pooled and analyzed by GC–MS. Figure 2 shows the total ionization plot (trace B) and the mass chromatogram of m/e 57 (trace C) for these combined fractions. The peak at scan 165 is elemental sulfur. Its mass spectrum shows peaks separated by 32 mass units and a molecular ion at m/e 256 corresponding to eight sulfur atoms. The presence of sulfur is not unexpected in such highly anoxic sediment (4).

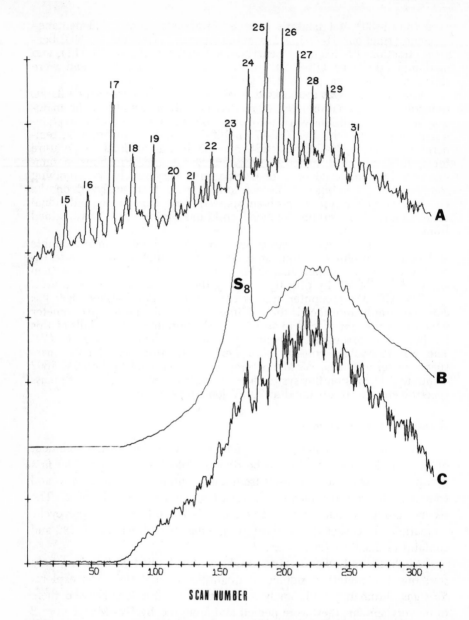

Figure 2. (A) *Mass chromatogram of m/e 99 representing the distribution of normal alkanes in Charles River water. The numbered peaks represent normal alkanes of the indicated chain length. Data are from Ref. 1. (B) Total ionization plot of fractions 4–7 from the extract of Charles River sediment. (C) Mass chromatogram of m/e 57 representing the distribution of alkanes in Charles River sediment.*

The other components of this mixture are not resolved from one another and form a wide continuum in the gas chromatogram centered at scan 220 which corresponds in retention time to octacosane. The general nature of these compounds can be determined by inspection of all the mass chromatograms from this mixture. From these data it can be concluded that these compounds are alkanes, cycloalkanes, and alkenes (containing up to three double bonds) with considerable branching. A mass chromatogram indicative of the alkanes found in the river water (at 2 m depth) is included for comparison (Figure 2A); the GC peaks caused by the individual normal alkanes are readily apparent. In the sediment, however, the normal alkanes are missing as noted by the lack of regularly spaced predominant peaks in the mass chromatogram of m/e 57 (Figure 2C). The lack of normal alkanes in the sediment is believed to result from biodegradation. The facile microbiological decomposition of normal alkanes (as opposed to branched species) is well known (5).

Polycyclic Aromatic Hydrocarbons. Gas chromatographic–mass spectrometric analysis of fractions 31–91 indicated the presence of various polycyclic aromatic hydrocarbons (PCAH) with molecular weights of 178–316. Because it is impossible to discuss in detail all the chromato-

Table I.   Polycyclic Aromatic Hydrocarbons Found in Sediment of the Charles River Basin by GC–MS

Compound Found	Maximum No. of Alkyl Carbon Atoms	Unsubstituted Compounds		
		Elemental Composition	No. of Rings + Double Bonds	MW
Anthracene or phenanthrene	12	$C_{14}H_{10}$	10	178
4,5-Methylenephenanthrene	9	$C_{15}H_{10}$	11	190
Pyrene (UV confirmation[a]) Fluoranthene (UV confirmation[a])	15	$C_{16}H_{10}$	12	202
Benzofluorenes (isomers unknown)	—	$C_{17}H_{12}$	12	216
Benzo[a]anthracene Chrysene	10	$C_{18}H_{12}$	13	228
Benzo[b or k]fluoranthene Benzo[a]pyrene Benzo[e]pyrene Perylene (trace quantities)	5	$C_{20}H_{12}$	15	252
$C_{22}H_{12}$ (isomer unknown)	3	$C_{22}H_{12}$	17	276
Dibenzopyrenes or dibenzofluoranthenes (isomers unknown)	3	$C_{24}H_{14}$	18	302

[a] A UV spectrum of fraction 43 showed peaks characteristic of a 1-1 mixture of pyrene and flouroanthene.

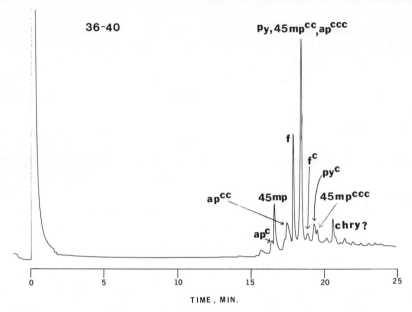

Figure 3. Gas chromatogram of fractions 36–40 from the extract of Charles River sediment. ap, anthracene or phenanthrene; 45mp, 4,5-methylenephenanthrene; f, fluoranthene; py, pyrene; chry, chrysene. The number of superscript c's indicates the number of alkyl substituted carbon atoms.

graphic and spectrometric evidence, only a few examples will be presented. A complete summary of all compounds found in these fractions is presented in Table I. For each compound listed, the mass spectrum and gas chromatographic retention times agreed with those of authentic material. The maximum number of alkyl carbon atoms found for the various polycyclic systems is also given in Table I. The exact substitution of these alkyl carbon atoms on the ring system is, of course, not known. It seems likely that many isomers are present simultaneously. The data on alkyl substitution are based on the mass spectra and relative GC retention times of the methyl and most dimethyl or ethyl derivatives and on peaks in the mass chromatograms corresponding to the higher alkyl homologs.

As an example of the chromatographic and mass spectrometric data for these aromatic fractions, Figure 3 shows the gas chromatogram of combined fractions 36–40. Examination of the mass spectra corresponding to the various chromatographic peaks, the gas chromatographic retention times, and the liquid chromatographic behavior allows one to make the structural assignments shown in Figure 3. Anthracenes or phenanthrenes containing one, two, and three carbon atoms as alkyl substituents are present. The two major components are fluoranthene and pyrene. Methyl fluoranthene and methyl pyrene were also identified. Mass chromatograms indicated the presence of higher alkyl substituted anthracenes or phenanthrenes, fluoranthenes, and pyrenes.

The mass spectrum of an interesting component in this mixture (*see* Figure 4) was identified as that of 4,5-methylenephenanthrene, the presence of the methylene group causing a characteristically abundant M − 1 ion at m/e 189. Alkyl substituted derivatives of this compound are also present in the combined fractions 36–40. In fact, the mass spectrum taken during the emergence of the largest chromatographic peak in this fraction (*see* Figure 5) clearly shows the C_2 substituted 4,5-methylenephenanthrene. Its molecular weight is 218 amu, and it exhibits fragment ions at m/e 203 and 189. In addition, this mass spectrum (Figure 5) shows the presence of pyrene (m/e 202) and the three-carbon atom substituted anthracene or phenanthrene, giving ions at m/e 220 and 205. This mass spectrum is an example of the emergence of three different ring systems in the same chromatographic peak; the coincidence of differing levels of alkyl substitution causes them to coelute. In a similar manner, other polycyclic aromatic ring systems and their alkyl substituted derivatives were identified in other fractions. Table I summarizes this information.

When taken together with the previous data on the aromatic hydrocarbons present in the water (*1*), it is now apparent that PCAH ranging from naphthalene to dibenzopyrene are present in the Charles River Basin system. The more water-soluble of these PCAH, the naphthalenes, are found in the water only; those of intermediate solubility, anthracenes or phenanthrenes, are found in both water and sediment; and those of

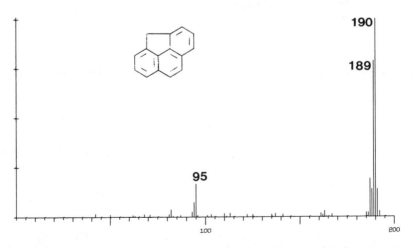

Figure 4. Mass spectrum corresponding to the peak labeled 45 mp in Figure 3. It has been interpreted as that of 4,5-methylenephenanthrene.

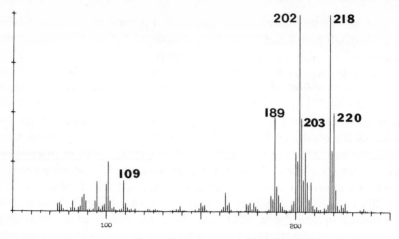

Figure 5. Mass spectrum corresponding to the most intense peak in Figure 3 (retention time = 18.5 min). See text for interpretation.

low solubility, molecular weights above 228, are found only in the sediment. The sources of these compounds will be discussed below.

Phthalate Esters. Fractions 92–122 contained large amounts of di-(2-ethylhexyl)phthalate and much smaller amounts of di-n-butyl phthalate and dioctyl adipate. The ubiquity of phthalate esters as laboratory contaminants is well known. In this case, however, the very high abundance of phthalates in these later fractions and their absence in previous

Table II.   Nonlinear Abundances of Polycyclic of the Charles River Basin

Total Carbon Number

14	15	16	17	18	19	20	21	22	23
13600	9676	7944	5208	3132	1920	1460	1732	2132	1812
6792	6172	7028	5044	3676	2724	2132	2124	1984	1860
2492	1248	16720	10468	7020	3932	2236	1500	1588	1104
0	76	6320	640	12444	7036	4232	2608	1316	696
0	532	1748	2164	9140	5404	4524	2996	1640	1484
0	0	0	88	2388	556	14408	7576	3744	1868
	0	0	460	288	168	5908	1568	8152	3528
		0	0	0	1356	3188	1768	12232	4344
			0	0	0	0	92	3016	436
				0	0	0	1072	928	88
					0	0	0	0	624
							0	0	60
								0	0

a These numbers are all divisible by four because a 14-bit A/D converter feeds a 16-bit computer word.

fractions and in various blank analyses indicate that the phthalates are indeed present in the Charles River sediment and are not artifacts. This should not be surprising since phthalate esters were among the more abundant components in the river water (*1*). It seems likely, therefore, that the river sediment accumulates these compounds from the water. Degradation by anaerobic bacteria, expected to be present in the sediment, could be slow and, thus, over a period of time high abundances of phthalates could accumulate.

High Resolution Mass Spectrometric Quantitation of Polycyclic Aromatic Hydrocarbons. Because of the striking abundance and variety of the PCAH found in the sediment by GC–MS, it was of interest to measure the relative abundance of these hydrocarbons. This was accomplished by HRMS. An aliquot of the total pentane-soluble fraction from the second batch of extracted sediment was introduced into a high resolution mass spectrometer through the direct introduction probe system and vaporized at a continually increasing temperature while seven exposures on a photographic plate were made. After development, the plate was read on the computerized comparator, and the exact masses with their intensities (in nonlinear arbitrary units) were stored on magnetic tape. Later these exact masses were converted to elemental compositions (*3*), and the corresponding intensities were then arranged as tables of carbon number *vs.* number of double bonds and rings. One such table for each of the seven exposures was generated, and then a composite

Aromatic Hydrocarbons Found in the Sediment as Determined by HRMS[a]

				Total Carbon Number			
24	*25*	*26*	*27*	*28*	*29*	*30*	Rings + Double Bonds
1580	1276	1032	916	292	256	64	10
1444	1052	924	716	292	152	100	11
1148	764	616	420	116	0	40	12
588	328	176	144	76	0	24	13
960	564	84	20	20	0	0	14
976	352	68	44	0	0	0	15
1536	424	132	24	0	0	0	16
2652	1032	268	44	28	0	0	17
8208	2384	940	188	0	0	0	18
4236	524	2432	536	20	0	0	19
1420	128	6596	1520	524	24	0	20
100	32	1104	88	2376	100	0	21
0	324	88	0	2240	0	36	22

table was formed by adding the corresponding entries in each of these tables. This final set of data is shown in Table II.

Because PCAH display mass spectra with extremely abundant molecular ions, this table can be thought of as a relative abundance display of the molecular ions for those PCAH present in the sediment sample. For example, the entry at 16 carbon atoms and 12 double bonds or rings corresponds to a mass of 202.0783 and represents the combined abundance of pyrene and fluoranthene (16,720 units). The combined abundance of all of the methyl derivatives of both these compounds and of the benzofluorenes is given by the entry immediately adjacent on the right, namely 10,468. In a similar fashion, all of the benzofluoroanthrenes and benzopyrenes are lumped together in one intensity measurement of 14,408 units (20 carbon atoms, 15 double bonds and rings). Unfortunately, the entries in this table are not a linear function of compound abundance; in fact, the exact functional relationship is not known. It is known, however, that the higher the entry in Table II, the more abundant are the compound or compounds represented by that entry. Thus, such a table is a crude semiquantitative indication of the relative abundance of the PCAH and their alkyl-substituted derivatives.

Despite this limitation one notices a number of features about the relative distribution of PCAH in Charles River sediment:

1. Pyrene and fluoranthene (m/e 202) are the most abundant group of isomers present; the $C_{20}H_{12}$ group is the second most abundant.

2. Coronene (C_{24}, 19 double bonds and rings) is present (4236 units), but it is less abundant than the dibenzopyrenes (8202 units).

3. An additional series of PCAH based on the elemental composition $C_{26}H_{14}$ and including the methyl and ethyl or dimethyl derivatives is present.

4. Within most series of homologs, the abundance monotonically decreases as substitution increases. In fact, a semilogarithmic plot of certain of these data (see Figure 6) shows a linear decrease in abundance as alkyl substitution increases within a given ring series. Indeed, the $C_{18}H_{12}$ series (molecular weight 228) displays a linearly decreasing abundance over 10 atoms.

The cause of this linearity is not yet clear. Nevertheless, displays of this sort are a particularly powerful, graphic means of semiquantitatively characterizing PCAH mixtures, and, very importantly, their alkyl substituted homologs.

Conclusions

The single most abundant class of compounds identified in the sediment of the Charles River Basin is the polycyclic hydrocarbons. This finding is unusual and surprising, and one immediately wonders about their source. [Concurrent work has shown that mixtures of PCAH of considerable complexity and which contain many of the same compounds listed in Table I are also present in a recent near-shore marine sediment (6)]. There are four possible sources of PCAH in the aqueous environment.

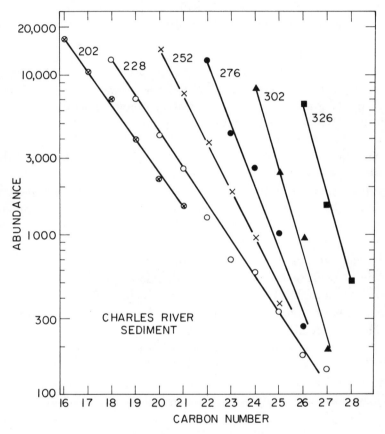

Figure 6.   Abundance (taken from Table II) of consecutive members for different PCAH homologous series vs. total number of carbon atoms in the molecule. The molecular weight of the unsubstituted homolog for each series is given adajacent to each line.

Petroleum. The presence of PCAH in petroleum is well known. Naphthalene, phenanthrene, fluorene, pyrene, chrysene, triphenylene, perylene, several benzanthracenes, and various alkyl substituted derivatives of these compounds have been found in crude petroleum (7). In addition, the PCAH content of those petroleum products which have been produced by thermal cracking is much greater than in the crude oil.

These petroleum-derived PCAH can enter the Charles River by numerous paths. The most dramatic and straightforward is dumping of waste motor oil into the urban runoff system. Another more indirect path is emission into the atmosphere of PCAH present in fossil fuels. For example, PCAH present in home heating oil or gasoline can be transported relatively unchanged through the combustion zone into the atmosphere where they will eventually enter the aqueous environment.

Combustion. The incomplete combustion of fuel of all sorts (petroleum, methane, coal, refuse, *etc.*) can produce PCAH by free radical reactions in the flame zone (8). Emission of these compounds into the air can occur both from mobile and stationary sources and is usually associated with soot production. PCAH from combustion sources could reach the Charles River mostly by way of rainwater which both scrubs them from the air and transports already precipitated PCAH (adsorbed on soot) by way of terrestrial runoff.

Biosynthesis. Although the data are somewhat preliminary, there are indications that PCAH can be produced by anaerobic bacteria and by some algae. In 1959 ZoBell (9) isolated 367 mg of "presumed aromatic compounds" from a 5-gal. culture of anaerobes. Specific compounds were not identified. *Clostridium putride* and *E. coli* seem to produce benzo[a]pyrene and other unidentified PCAH (10). The fresh water alga *Chlorella vulgaris* may synthesize several PCAH from ^{14}C-labeled acetate (11). It seems quite possible, therefore, that a mixed population of various anaerobic bacteria and algae, which would be present in an anoxic body of water such as the Charles River and its sediment, could produce considerable quantities of PCAH.

Chemical Synthesis. If cholesterol is heated with elemental sulfur at 150°C for several days, a mixture of PCAH results (12), the detailed composition of which is not known. Other dehydrogenation reagents, such as selenium, catalyze the production of alkyl naphthalenes and phenanthrenes from naturally occurring terpenes (13). Thus, in principle, it is possible that the PCAH in an anoxic sediment could result from the chemical dehydrogenation of various naturally occurring compounds. These reactions may take place slowly over the years and may be catalyzed by trace elements, elemental sulfur, and clay minerals.

Without detailed information on the characteristics of the PCAH mixtures produced by each of these four sources, it is difficult to specify which may be the most predominant in the Charles River system. It seems likely, however, that all of these possibilities play a role and that the PCAH which have been observed are the result of both natural processes and human activity.

Acknowledgments

W. G. Biemann thanks the Deutscher Akademischer Austauschdienst for a fellowship during 1972 and 1973. The instrumentation was supported (in part) by National Institute of Health Research Grant RR00317 from the Division of Research Facilities and Resoruces.

Literature Cited

1. Hites, R. A., Biemann, K., *Science* (1972) **178**, 158.
2. Hites, R. A., Biemann, K., *Anal. Chem.* (1967) **39**, 965; (1968) **40**, 1217; (1970) **42**, 855.
3. Biemann, K., in "Applications of Computer Techniques in Chemical Research," p. 5, Institute of Petroleum, London, 1972.
4. Kellogg, W. W., Cadle, R. D., Allen, E. R., Lazrus, A. L., Martell, E. A., *Science* (1972) **175**, 587.
5. Dugan, P. R., "Biochemical Ecology of Water Pollution," p. 90, Plenum, New York, 1972.
6. Giger, W., Blumer, M., *Anal. Chem.* (1974) **46**, 1663.
7. Speers, G. C., Whitehead, E. V., "Organic Geochemistry," G. Eglinton and M. T. J. Murphy, Eds., p. 638, Springer-Verlag, New York, 1969.
8. Badger, G. M., *Natl. Cancer Inst. Monogr.* (1962) **9**, 1.
9. ZoBell, C. E., *N. Z. Oceanogr. Inst. Mems.* (1959) **3**, 39.
10. ZoBell, C. E., *Proc. Jt. Conf. Prev. Control Oil Spills* (1971), 441.
11. Borneff, J., Selenka, F., Kunte, H., Maximos, A., *Environ. Res.* (1968) **2**, 22.
12. Douglas, A. G., Mair, B. J., *Science* (1965) **147**, 499.
13. Mair, B. J., *Geochim. Cosmochim. Acta* (1964) **28**, 1303.
14. Smith, J. D., Process Research Inc., Cambridge Mass., 1970, unpublished data.

RECEIVED January 3, 1975

17

Occurrence and Implication of Sedimentary Fluorite in Tampa Bay, Fla.

DEAN F. MARTIN and WILLIAM H. TAFT

University of South Florida, Tampa, Fla.

Phosphate processing caused large amounts of gaseous fluoride to be emitted as a waste product through stacks. Scrubbers were installed by a plant located on Tampa Bay, Fla., and the collected fluoride was then discharged into the Bay. In July 1973, the daily discharge of fluoride entering the Bay was approximately 24,000 lb. A study of water and sediment samples in Tampa Bay adjacent to a discharge canal of the plant revealed (1) a deltaic deposit of sedimentary fluorite, (2) remarkably low pH meter readings (3.3) that indicated the buffer capacity of $50 \times 10^6 \, m^3$ of estuarine water was virtually exhausted, (3) fluoride concentrations as much as 40 times the concentration in normal seawater, and (4) temperature differentials in relatively shallow water (0.6 m) that were the reverse of what would normally be expected.

The geochemistry of fluoride in estuarine waters has received increasing attention (*1, 2, 3, 4, 5*) because of concern about the primary sources of fluoride in continental waters—rainwater of ultimate marine origin, volcanic emanations coupled with atmospheric precipitation, rock weathering, and industrial. Generally, rock weathering is considered least as a primary source of fluoride. Kilham and Hecky (*6*) demonstrated the significance of weathering of fluoride-rich volcanic rocks in East Africa. The Africa site is far removed from industrialized regions and may provide the best area for studying the preindustrial fluoride cycle and estimating the impact of man.

Tampa Bay, Fla., may well represent another area that is ideal for studying the impact of man on the fluoride cycle, particularly in an estuarine environment. Indeed, we believe that the present study provides the basis for studying a situation at what may be the ecologically worst stage, followed by improvement.

In Florida, fluoride and phosphate are chemically and industrially associated. Phosphate deposits in Florida occur as sedimentary phosphorite of Miocene age (10–15 million years old). The principal mineral is apatite, $Ca_5(PO_4)_3F$, containing about 4% fluorine. The deposits are centered in a 500-sq mile area around Bartow. Mining was initiated in 1890, and in 1972, Florida produced more than 30 million tons of phosphate valued at about $170 million. Florida supplies over three-fourths of United States needs and roughly one-third of the world needs (7).

The processing of this resource underwent a number of changes as environmental degradation was noted and new technology became available to resolve those problems. According to Hendrickson (8) as much as eight tons of gaseous hydrogen fluoride per day was discharged into the atmosphere within an area of 100 sq miles as a result of treatment of the ores with sulfuric and phosphoric acid. That time has long passed, and phosphate processing companies have assumed considerable ecological responsibility. For example, complaints by local residents, orange growers, cattlemen, and commercial florists led one such company, Gardinier, Inc. (then owned by U.S. Phosphoric) to install scrubbers in about 1963. Rather than being released into the air, collected fluoride was discharged into Tampa Bay. The amount of fluoride discharged into the Bay varied with the amount of phosphate produced, but in July of 1973, the daily discharge amounted to about 24,000 lb. In addition, the effluent contained about 27,000 lb of phosphate and 3,000 lb of nitrate per day.

This paper describes results of analyses of Tampa Bay water near the Gardinier plant in July 1973 and July 1974 before and after reductions in amounts of pollutants had been promised. Clearly, the discharge of 24,000 lb of fluoride on a daily basis seems excessive, and in July 1973 when the first sampling took place, Gardinier had hoped to reduce the discharges by 97% within six months. The data collected in July 1973 serve as a benchmark against which to measure the results of pollution abatement, and the data collected in July 1974 measure the progress of environmental restoration.

Physical Observations

The location of Gardinier's plant in relation to Tampa Bay is indicated in Figure 1. Two canals discharge processing wastes into the Bay. A deltaic deposit of fluorite that may be the only such deposit of sedimentary fluorite known in the world is located at the mouth of both canals. In cross-section, the fluorite deposit is about 3 in. thick at the

*Figure 1.   Study site on Hillsborough Bay portion
of Tampa Bay*

initial discharge point and rapidly diminishes to translucent flakes at the
outer edges of the deposit, roughly 350 m into the Bay.

Temperature and pH data were obtained at the time of sampling.
Temperatures were obtained using the same mercury thermometer in
1973 and 1974. In July 1973, the temperature increased rapidly during
equilibration while water temperature was read at the surface (30-sec
period). Thus, the temperatures in Table I should be regarded as minimum values. The thermal problem was not noticeable during the July
1974 sampling. All pH values were obtained in the field with a Beckman
model G pH meter, standardized with pH 7 buffers. The field values
were checked again in the laboratory and generally agreed within 0.2
units.

Salinity was obtained using the Harvey method (9) and was checked
with a hand-held refractometer (9). Silicate and phosphate analysis were
obtained by a Technicon AutoAnalyzer II using standard procedures
described in the EPA Methods Manual (10).

Fluoride was determined mainly by using an Orion fluoride electrode
(model 90-01) with a Corning model 10 expanded pH meter. The procedures described earlier (9, 11, 12) were modified as follows. To a 10 ml
sample of water in a plastic beaker was added 40 ml of stock seawater
(30‰), followed by 10 ml of total ionic strength buffer (TISB, pH = 5.5,
ionic strength 1.9) (13). Initial reading (= E_x) was recorded after

constancy was obtained (usually less than 5 min). Readings ($= E_x$) were also recorded after addition of stock fluoride solution (1000 ppm F^-) in 0.1-ml increments (or addition of 2, 4, 6, 8, 10 ppm F^-). The data were plotted on semi-logarithmic paper: E_x (in mV) as a function of log ppm added fluoride. The linear relationship was extrapolated to obtain the value of apparent initial fluoride, $(F^-)_i$, knowing E_i. The apparent fluoride concentration $(F^-)_o$ in the original sample was calculated using Equation 1:

$$(F^-, \text{ppm})_o = \frac{(F^-)_i - (F^-)_s}{\text{D.F.}} \tag{1}$$

where $(F^-)_s$ is the apparent fluoride concentration in the stock seawater (a value obtained by using 10 ml of distilled water instead of sample),

Table I.   Physical and Chemical Properties of Tampa Bay Water Samples

Station	Salinity (‰)	t (°C)	pH [a]	PO_4–P (ppm)	SiO_2–Si (ppm)	F^- (ppm)	Distance from Discharge Canal (m)
			1973				
1B [b]	27.8±0.3	36	3.3	36.7	44.1	43.0	15
2S	21.1±0.1	28	3.95	32	18.7	23.3	155
2B	27.1±0.3	34	3.3	33	46.1	34.8	155
3S	21.1±0.1	28	5.70	20.1	10.3	18.5	305
3B	22.6±	28.5	4.60	22.5	17.1	14.0	305
4S	31.1±0.1	30	3.8	31.3	19.0	21.5	115
4B	29.6±0.3	34	3.2	35.5	46.9	30.8	115
5S	23.8±0.2	33	3.5	35.4	25.5	36.5	155
5B	31.7±	35	3.2	33.6	49.2	25	155
6S	22.4±0.4	33	3.6	35.5	25.6	34.2	305
6B	31.9±3.5	33	3.2	40.3	34.8	35.0	305
7S	21.2±0.1	28.5	5.5	23	10.7	16.3	265
7B	34.8±0.6	31	3.4	33.7	40.0	36.3	265
			1974				
SC [c]	30	33	6.68		5.0	2.1	0
2S	15	28	6.90		1.4	2.6	15
3S	13	29	7.22		1.6	2.6	32
5B	16	31	7.21		1.1	2.0	265
7S	15	30	7.60		0.8	1.8	425
8 GFD [c]	32	32	2.28		61.2	330±6 [e]	0
						346±20 [f]	

[a] pH meter reading, B.
[b] S, surface; B, bottom (Ca. 0.6 m at time of collection, August 23, 1973).
[c] SC, southernmost canal.
[d] Sample collected near gypsum field ditch (*see* Figure 2).
[e] Fluoride ion electrode method, internal standard, dilutions at 0.1, 0.02, 0.01.
[f] Colorimetric method with internal standard, dilution at 0.004.

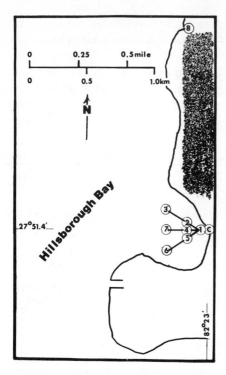

Figure 2. Location of specific sampling sites in relation to southern and northern discharge canals

$(F^-)_i$ is the apparent initial fluoride in the TISB-diluted sample, and D. F. is the dilution factor, 0.2.

The accuracy of the method was indicated by the value of $(F^-)_s$, which was 1.05 ppm, and which would correspond to a value of 1.3 ppm for the sample of salinity, $S = 35\%o$. For comparison, the calculated value for salinity of $35\%o$ would be 1.28, based upon the reported value for standard seawater (*13*). The precision was estimated for samples 6S and 6B for which the mean and standard deviations were 34.2 ± 0.8 and 35.0 ± 0.3, respectively; the corresponding relative deviations were 2.3 and 0.9%. There was no significant variation in fluoride values during the 24 hr after being stored in a plastic container and refrigerated at < 4°C. In addition, fluoride was also determined for one unique sample colorimetrically, using an lanthanum–alizarin complexone reagent (*14*). Data are compared in Table I.

Observations of Organisms in the Sampling Area and Environmental Restoration

The contrast in the organism observations before and after fluoride emission abatement was dramatic. During July 1973, pH-meter readings that were remarkably low for estuarine samples were observed within the area of fluoride deposition (Table I), and it is not remarkable that there were no visible signs of life within the area. For example, an abundance of jellyfish were living in Tampa Bay during the field study, but in the study area the four to six such organisms that were seen were all dead. We presume that the jellyfish had been killed by the chemical discharge. In addition, no barnacles were present in the area. In contrast, the bottom sediment during July 1974 had a normal distribution of marine worms and barnacles attached to sticks and sedimentary fluorite typical of Tampa Bay, and, in addition, fish were observed. The existence of the normal compliment of marine worms indicates a bottom sediment that is beginning ecological restoration. Most of the barnacles were dead which suggested a period of much less pollution discharge since July 1973 and then an increase that destroyed the animals. At the time of the 1974 collections, it was not possible to determine the precise state of restoration.

It was expected that sampling in 1974 would provide low fluoride concentrations and high pH values, thereby indicating a return to a more estuarine condition. However, as can be seen by Table I, the 1974 results have much higher fluoride and much lower pH-meter reading values than those of 1973. The company (Gardinier) explained the 1974 values by excessive rainfall and runoff from their storage facility. Regardless of the reason, concentrations of fluoride in Tampa Bay at the 1974 collection were excessively high and reflected a continuation of fluoride pollution in Tampa Bay six months after Gardinier promised to obey a temporary operating permit that called for a 97% reduction by January 1, 1974.

Conclusions

Several unique features emerged from this study. The most obvious was the existence of sedimentary fluorite. This deposit is considered unique in the world. It is a key and significant feature that the amount of fluorite is not greater, and this argues for considerable complexing of the fluoride, presumably as H_2SiF_6. This compound results when silicon tetrafluoride, obtained during the treatment of phosphatic rocks (Equations 2 and 3) is washed out of the waste gases by water and hydrolyzed (Equation 4).

$$Ca_3(PO_4)_3F + 5 \ H_2SO_4 \rightarrow 5 \ CaSO_4 + 3 \ H_3PO_4 + HF \qquad (2)$$

$$4 \ HF + SiO_2 \rightarrow SiF_4 + 2 \ H_2O \qquad (3)$$

$$3 \ SiF_4 + 2 \ H_2O \rightarrow SiO_2 + 2 \ H_2SiF_6 \qquad (4)$$

Study of the mass balance of hydrogen, silicon, and fluorine found in the bottom and surface samples should indicate the persistence of the complex. For near-bottom samples, for example, the effluent appears to be silicon rich, and the agreement between the amount of H_2SiF_6 calculated on the basis of $[H^+]$ and $[F^-]$ is fairly good while agreement for surface samples is poorer. The relationship between $[H_2SiF_6]_F$ and $[H_2SiF_6]_H$, the amount of fluorosilicic acid calculated on the basis of fluoride and hydrogen-ion analyses, respectively, is plotted (Figure 3).

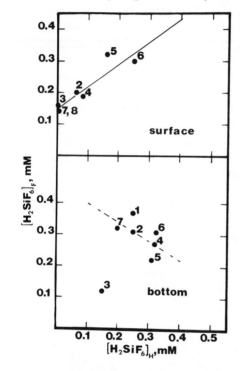

Figure 3. Concentration of [H_2-SiF_6] in the sampling area (July 1973)—F-limiting basis as a function of H-limiting basis

For the bottom samples, the agreement between two limiting values is good, and there appears to be little excess fluoride over what would be calculated for H_2SiF_6. The correlation between the two values for surface samples is good ($r = 0.928$, $P \leqslant 0.001$) and statistically significant, but the slope suggests that fluoride is in considerable excess over hydrogen ion, and the logical suggestion is that the disassociation of the SiF_6^{2-} species occurs during mixing with estuarine water. Silicon is clearly in excess over that required for complex formation, and the excess of silicon relative to fluoride is greater for bottom samples. Complexing by hydrogen ion is also a considerable possibility at these pH values, though the results in Table I appear better explained by complexing as H_2SiF_6.

A second unique feature is the pH-meter readings, which were remarkably low (3.3) for several stations in July 1973. The values indicate that the buffering capacity [ca. 2.5 meq/kg, but see Sillén (15) for another view] of 50×10^6 cu m of estuarine water was exhausted. In contrast, the same stations when occupied in July 1974 had returned to a more natural pH (6–7). On the other hand, we were startled by the low pH (3.3) at a station 150 ft opposite the northernmost discharge canal in July 1974 (*see* Figure 2). A sample of the water from that canal had the lowest pH-meter reading that we have yet encountered (2.2–2.3), and this sample of water was exceptionally well buffered because the pH-meter reading remained constant for several weeks when the sample was stored in the refrigerator at $< 4°C$. Discharge may have accidentally occurred during the July 1974 sampling visit although data provided by another agency (Hillsborough County Environmental Protection Agency) suggest that the problem persisted into August 1974.

This view is supported by another unique feature, the excessively high fluoride concentrations in the north discharge canal. It appears the concentration of fluoride was about 340 ppm. This is a value that was consistently obtained by two methods (specific ion electrode and colorimetric analysis), but a 100-fold dilution was required for the first method and a 250-fold one for the second. We recognize that the problems of errors are magnified by dilutions of this magnitude, but it is evident that the fluoride concentrations were unusually excessive.

Finally, a fourth observation, the temperature differential, was observed during the first sampling visit, although not during the second. The temperature differentials in shallow water (about 0.6 m) were not only significant but were the reverse of what might reasonably be expected. As noted earlier, the temperature increase at the base of the water column indicated a thermal pollution problem in July 1973. Temperature differentials of as much as $6°C$ between surface and bottom water temperatures in such shallow waters indicate the pollutant discharge did not mix rapidly with the overlying water.

The present study indicates the use of a fluoride specific ion electrode as a means of measuring fluoride as a specific pollutant, but more to the point, the study indicates the possibility of environmental restoration when specific emissions are curtailed. If the abatement program is continued, apart from occasional accidental lapses, a significant result of the unique deltaic sedimentary fluorite deposit may be anticipated—fishing should be better there than in many places of Tampa Bay where absence of a firm bottom contributes to turbidity problems during storms and prevents benthic plants from using a useful niche.

Acknowledgment

We are grateful for the technical information and assistance provided by Donald Graff, manager of Gardinier, Inc.; Robert Knight of the

Florida Marine Patrol; Val Maynard, U.S.F. Department of Marine Science, who performed the silicate and phosphate analyses; James Goetz, who performed the x-ray analyses; and Joseph Simon, who assisted in identification of marine worms and other organisms. Finally, we are grateful for the cooperation of the Hillsborough County Environmental Protection Agency.

Literature Cited

1. Brewer, P. G., Spencer, D. W., Wilkness, P. E., *Deep-Sea Res.* (1970) **17**, 1.
2. Carpenter, R., *Geochim. Cosmochim. Acta* (1969) **33**, 1153.
3. Kilham, P., Hecky, R. E., *Limnol. Oceanogr.* (1973) **18**, 932.
4. Shiskina, O. V., *Geochim. Int.* (1966) **3**, 152.
5. Windom, H. L., *Limnol. Oceanogr.* (1971) **16**, 806.
6. Livingstone, D. A., *U.S. Geol. Surv. Prof. Pap.*, 1963, 440-G.
7. Anon, *Fla. Conserv. News* (August, 1973) 3.
8. Hendrickson, E. R., *J. Air Pollut. Control Ass.* (1961) **11**, 220.
9. Martin, D. F., "Marine Chemistry," vol. 1, 2nd ed., Dekker, New York, 1972.
10. Environmental Protection Agency, "EPA Methods for Chemical Analysis of Water and Wastes," U. S. Department of the Interior, Washington, 1971.
11. Taft, W. H., Martin, D. F., *Environ. Lett.* (1974) **6**, 167.
12. Warner, T. B., *Science* (1969) **165**, 178.
13. Riley, J. P., Chester, R., "Introduction to Marine Chemistry," p. 81, Academic, New York, 1971.
14. Greenhalgh, R., Riley, J. P., *Anal. Chim. Acta* (1961) **25**, 179.
15. Sillén, L. G., *Science* (1967) **156**, 1189.

RECEIVED January 3, 1975. This work was supported partially by Public Health Service Research Career Development Award (KO4-GM 42569-05) from the National Institute of General Medical Sciences.

Changes in the Physical and Chemical Properties of Biogenic Silica from the Central Equatorial Pacific

I. Solubility, Specific Surface Area, and Solution Rate Constants of Acid-Cleaned Samples

DAVID C. HURD and FRITZ THEYER

Hawaii Institute of Geophysics, Honolulu, Hawaii 96822

General trends of decreasing solubility of acid-cleaned radiolarians with increasing age suggest that cherts and porcelanites (recrystallized cristobalite and quartz) are presently forming. The thermodynamic properties of biogenic silica are between those of silica gel and cristobalite. The specific surface area of biogenic silica assemblages has decreased by two orders of magnitude in the last 40 million years. Heterogeneous solution rate constants for pure substances yield valuable information regarding the free energy of activation of solution processes. These constants are quite sensitive to contamination from a mixture of various silica forms and may not be as immediately useful as the solubility information.

Biogenically precipitated silica is a metastable silica polymorph which must eventually alter to quartz under the earth's surface conditions. Present observations of deep-sea sediments suggest that this transformation may occur directly or through an intermediate, alpha cristobalite. Several models have been proposed to ascertain the rate at which these processes occur. This series of papers tests these models and offers simple but powerful methods for detecting changes in crystal form as a function of geologic age.

As noted in the title, the first section of this research deals with the changes in solubility, specific surface area, and solution rate constants of the substances studied. Since it is important to understand the extent to which these properties change with changes in crystal form, we first review previous workers' data on each topic. In addition, we consider the initial properties of biogenically precipitated silica relative to its more stable polymorphs and how these properties change as the transitions to these polymorphs occur. From the beginning it may be argued that the study of only acid-cleaned materials can hardly be extrapolated to the complex interactions occurring among silica polymorphs, metal oxides, and alumino–silicate minerals in the sediments. However, we suggest that such reactions alter only the rate and not the final outcome of the silica transitions. Further, if the properties of the starting material relative to the end products are not well understood, how is it possible to understand the extent and nature of these transitions?

Solubility of Biogenically Precipitated Silica, Vitreous Silica and Silica Gel, Cristobalite, and Quartz in Aqueous Solutions

A number of authors (1, 2, 3, 4, 5) studied the solubility of biogenically precipitated silica. The solubility of artificially precipitated silica (silica gel) and of vitreous silica, two forms of silica which have similar solubilities also have been studied (6–16). Biogenic silica and silica gel probably resemble each other more than either resembles vitreous silica. Sosman (17) and Iler (8) presented excellent discussions of silica gel and vitreous silica properties. Depending on the preparation method, degree of internal ordering, and specific surface area, the solubilities of vitreous silica and silica gel vary widely. However, both substances share similar ranges which are at least one order of magnitude greater than quartz in the 0°–25°C range of temperatures considered here. For this reason alone the two were lumped together in the present discussion. Equilibrium values for the two in seawater, pH 7.5–8.3, are in the range 1500–2000 μM at 25° $\pm$ 1°C.

The equilibrium solubility for low or alpha cristobalite in distilled water at 25°C (by extrapolation from higher temperatures) is ca. 250 μM (18). In a later paper (19) these authors showed that this material was saturated sevenfold in distilled water at room temperatures and that the above extrapolated value was not attained during the experiment (4.5 years). Although we believe that the extrapolated value is valid, it

Table I.   Values of Thermodynamic Properties

	Quartz[b]		Low Cristobalite[c]	
	5°C	*25°C*	*5°C*	*25°C*
$\Delta G°$	5.47 ± .17	5.43 ± .17	4.58 ± .2	4.58 ± .2
$\Delta H°$	6.0 ± .15	6.0 ± .15	4.58 ± .3	4.58 ± .3
$\Delta S°$	1.91 ± .02	1.91 ± .02	~0	~0

[a] The signs of all values are with respect to the reaction: (solid) → silica monomer.
[b] From measurements of Morey et al. (19)
[c] From measurements of Fournier and Rowe (18)

is clear that it is difficult to attain equilibrium in distilled water at room temperature. In his most recent paper, Fournier (*20*) gives high temperature (165°–250°C) solubilities for a high or beta cristobalite sample, which give lower temperature values (by extrapolation) of *ca.* 1350 μM at 25°C and 830 μM at 5°C. Stöber (*15*) did several experiments in 0.154N sodium chloride solutions at 25°C and obtained values intermediate to the above-mentioned low and high cristobalite data. However, his sample was not well defined mineralogically, and his results are therefore questionable. No data were found for seawater solubilities.

Equilibrium solubilities for quartz in distilled water at 25°C (also obtained by extrapolation from higher temperatures) are in the range 100–200 μM (*14, 19, 21, 22*); at 5°C by the same process, *ca.* 80–120 μM. The seawater solubility value of quartz at a slightly lower temperature (73 $\pm$ 5 μM at 20°C), obtained recently by Mackenzie and Gees (*23*), suggests that the data of Morey *et al.* (*19*) are the most reliable. The latter's estimates are used in this paper.

That each of the above forms (possibly excepting cristobalite in distilled water) has a reasonably well defined and reproducible equilibrium value suggests that the following familiar equations may be used to describe the net energy changes involved on reaching equilibrium (*24*):

Change in free energy (kcal/mole) $\Delta G° = -RT\ln(K_{eq})$ (1)

Change in enthalpy (kcal/mole) $\Delta H° = \dfrac{Rd\ (\ln K_{eq})}{d(1/T_{abs})}$ (2)

Change in entropy (cal/deg K/mole) $\Delta S_T = (\Delta H° - \Delta G°)/(T_{abs})$ (3)

Thermodynamic values at 5° and 25°C for each of the three forms of silica are given in Table I. The large differences in these values suggest that the changes in these properties as a function of the stability of the crystal structure should help to identify form changes in the sediment.

The most obvious change is that of solubility. Simply by cleaning the sample to remove clay minerals and absorbed cations and mixing the sample with seawater and allowing sufficient time for equilibration,

as a Function of Temperature and Form [a]

High Cristobalite[d]		Biogenic Silica[e]		Vitreous Silica, Silica Gel[f]	
5 °C	*25°C*	*5°C*	*25°C*	*5°C*	*25°C*
3.92 $\pm$ .2	3.92 $\pm$ .2	3.83 $\pm$ .03	3.81 $\pm$ .03	3.70 $\pm$ .15	3.73 $\pm$ .15
3.64 $\pm$ .15	3.64 $\pm$ .15	3.97 $\pm$ .58	3.97 $\pm$ .58	3.30 $\pm$ .03	3.30 $\pm$ .03
−0.96 $\pm$ .35	−0.96 $\pm$ .35	∼0	∼0	−1.44 $\pm$ .45	−1.44 $\pm$ .45

[d] From measurements of Fournier (*20*)
[e] From data in Appendix (this report) 1.0–6.4 × 10[6] years before present
[f] From review of Alexander, Krauskopf, and Siever by Wollast (*35*)

gross changes may be recognized easily in the surface properties of the biogenic assemblage *per se*. Changes in heat of dissolution are generally within 10% of the absolute values of the free energy changes, but alter at a slightly different rate producing an interesting entropy effect. There is a net gain in entropy when quartz dissolves and a net loss in entropy when either vitreous silica or silica gel dissolves. That is, not only are greater heats of dissolution required to remove a silica molecule from increasingly more stable crystal structures, but the degree of disorder of the hydrated monomer relative to the molecule in the more stable crystal structure increases as well.

The biogenic silica values were obtained from Table III for the time period Recent to 6.4 millions of years before Present (mybp). They show a net free energy change between that of cristobalite and the silica gel–vitreous silica combination as well as a similar intermediate status of enthalpy and entropy values. We suggest that careful characterization of the solubility of a biogenic silica sample at several temperatures may yield useful information regarding its transformation to a more stable substance. Caution must be used in interpreting these changes in solubility *per se* since:

1. Any given assemblage contains on the order of three to five dozen different species of radiolarians, diatoms, and sponge spicules. Preliminary investigations based on the refractive index of each species (to be elaborated upon in a subsequent report) suggests that almost every species is mineralogically slightly different from every other in a given assemblage.

2. The range of specific surface areas from Recent radiolarians to sponge spicules may vary by nearly three orders of magnitude. Since Alexander (25) has shown that for a series of silica gel sols, solubility varies as a function of both specific surface area and internal structure, only general trends in solubility can be discussed, and those, conservatively.

3. Assuming that this process occurs by dissolution of the more soluble phase and precipitation of the less soluble ones, a relatively thin veneer of less soluble material may well coat those species which are mineralogically more stable, further tending to preserve them at the expense of the less stable ones. While the bulk of the assemblage may be thus coated, a smaller percentage may slowly still yield high solubilities given long time intervals. This is further discussed in the section on dissolution rates.

Figures 1 and 2 show the change in solubility of the acid-cleaned radiolarian and sponge spicule assemblages at $26° \pm 1°C$ and $3° \pm 1°C$ as a function of geologic age. Also shown are the estimated values of the high and low cristobalite and low quartz solubilities at these two temperatures. The open and crossed circles represent the initial leveling off of the concentration of $Si(OH)_4$ *vs.* time curves, and the dots are the values reached after three to six months of constant agitation. Although there is some scatter in the data, there are clear trends of decreasing solubility with increasing sample age. At least two types of behavior are apparent: a gradual decrease in solubility with increasing age, suggesting at least by 60 ± 10 mybp cristobalite solubilities will be reached and one in which the process appears to be accelerated by a factor of three to four, and quartz solubilities are approached after only 15–20 mybp.

Figure 3 shows the approximate age *vs.* number of occurrences of recrystallized cristobalite (porcelanite) and quartz (chert) found at selected sites of the Deep-Sea Drilling Project. The number within each box gives the site at which the mineral was found in abundance. The age range of maximum occurrence of recrystallized silica forms (35–65 mybp) agrees quite well with the solubility trends shown in Figures 1 and 2.

Work by Harder (26) suggests that quartz, in the presence of various metal hyroxides at pH 7 between 5° to 80°C, spontaneously pre-

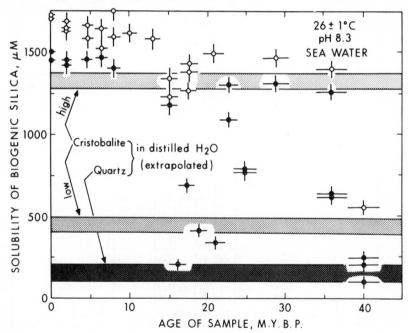

Figure 1.   Solubility of acid-cleaned biogenic silica in pH 8.3 seawater at 26° ± 1°C as a function of sample age.

Samples from core S-68-24 are marked with ⊕; many of these samples showed low solubilities at much earlier ages than the other cores. ● give dissolved silica concentrations for these same samples after an additional three to six months of constant agitation.

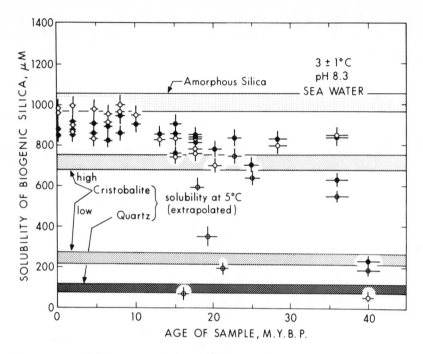

Figure 2.   Solubility of acid-cleaned biogenic silica in pH 8.3 seawater at 3° ± 1°C as a function of the age of the sample (symbols same as in Figure 1)

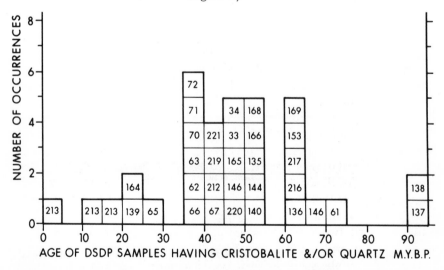

Figure 3.   Number of occurrences of recrystallized cristobalite or quartz from selected Deep-Sea Drilling Sites as a function of the age of the sample. Numbers within boxes refer to drilling sites. While the listing is not exhaustive, we do feel that it is representative.

cipitated in only a few weeks time from solutions which were much less than saturated with respect to silica gel. Since all deep-sea sediments which were squeezed at their *in situ* temperatures showed dissolved silica concentrations of not greater than 60% of amorphous silica saturation at those temperatures, it is puzzling why quartz does not actually form more rapidly.

Solution Rate Constants of the Various Forms of Silica

The first order congruent solution of amorphous silica is described by the equation

$$dC/dt = k_2(C_{sat} - C_{sol})S \qquad (4)$$

where k_2 is the first order rate constant in cm sec^{-1}, C_{sat}, the concentration of a solution saturated at a particular temperature and pH, and C_{sol} the solution being observed at time t in moles cm^{-3}, and S is the surface area of the solid per unit volume of solution in cm^{-1} (*1, 2, 22, 27, 28*). There are a number of concepts to consider when using such a formula to describe the dissolution of biogenic silica.

The importance of the surface area per unit volume term, S, cannot be overstressed. In the past virtually all investigators have lumped the k_2 and S terms together, without knowing what the S term was. This then generates a countably infinite number of dissolution constants and dC/dt combinations, none of which can be compared with another (*4, 5, 10, 28, 29, 30*). The importance of the S term, then, is that the k_2 value for the same substance under the same conditions of temperature, pH, and ionic strength is the same irrespective of the amount suspended in solution. Only by knowing the S term can the surface properties of spines or shells from different organisms be compared since this allows calculation of k_2 for a given set of conditions.

Figure 4 shows the change in specific surface area of the acid-cleaned assemblages with increasing age and reinforces the importance of determining the specific surface area of the solids involved. Even if no change in crystal form occurred during a 40 million year period, a comparison of the dissolution rate of equal weights of sample (assuming incorrectly that their surface areas were the same) would show a difference in initial solution rates of *ca.* a factor of 100. It is suggested that the observed two-orders-of-magnitude decrease results from both diagenetic and morphological changes. Although we are primarily concerned with these numbers insofar as they allow us to calculate S for a given experiment, they do allow us to quantitatively describe earlier micropaleontological observations relating to test structure such as "fragile" and "robust." Figures 5–10 show the degree of variation in geometrical

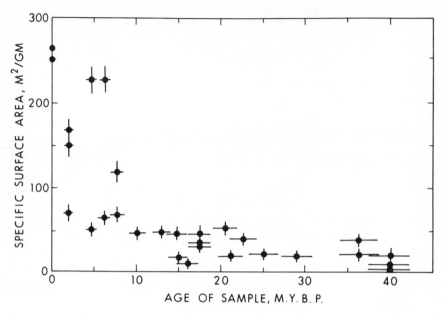

Figure 4.   Change in specific surface area in m²gm⁻¹ of acid-cleaned biogenic silica as a function of sample age

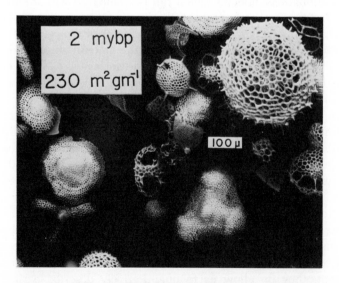

Figure 5.   Pleistocene age radiolarian assemblage having 230 m² gm⁻¹ specific surface area. Width of smaller white box is 100 μ; the resulting magnification is ca. 92×.

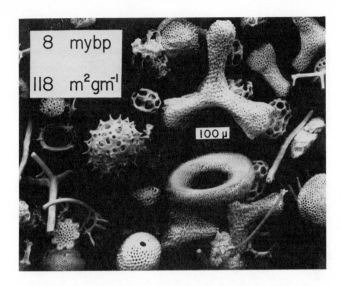

Figure 6.   Late Miocene age radiolarian assemblage having 118 m² gm⁻¹ specific surface area. Hollow branching rod on the left is a sponge spicule.

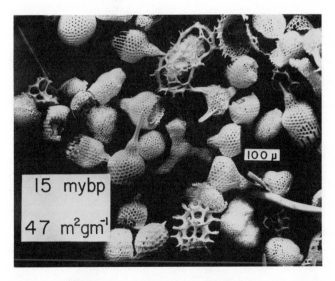

Figure 7.   Middle Miocene age radiolarian assemblage having 47 m² gm⁻¹ specific surface area

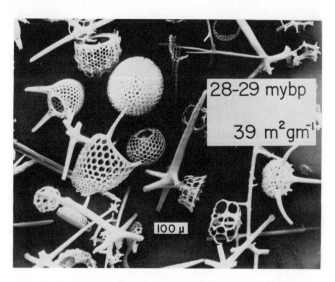

Figure 8. Late Oligocene age radiolarian and sponge spicule assemblage having 39 m² gm⁻¹ specific surface area

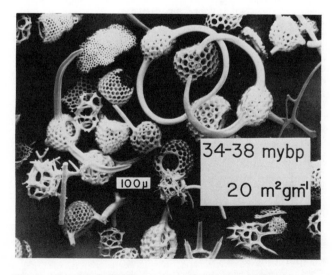

Figure 9. Early Oligocene age radiolarian assemblage having 20 m² gm⁻¹ specific surface area

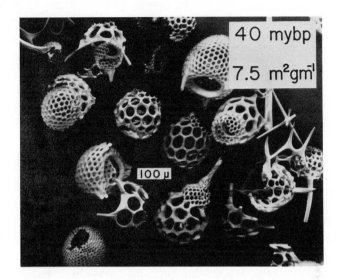

*Figure 10.   Late Eocene age radiolarian assemblage hav-
ing 7.5 m^2 gm^{-1} specific surface area*

complexity from one assemblage to the next as a function of the geologic
age of the sample.

The depolymerization rate constant k_1, expressed in moles cm^{-2} sec^{-1},
is the product of C_{sat} (or K_{eq}) and k_2 (*1, 2, 22, 27*). The chemical
meaning of such a constant is that for a given temperature, pH, and
ionic strength, k_1 represents the maximum solution flux per unit area
which can be expected from a given silica sample. At equilibrium this
flux must be equal and opposite to the product of k_2 and C_{sol} when $C_{sol} =
C_{sat}$; *i.e.*, $dC/dt = 0$.

The term dissolution rate also needs clarification. The dissolution
rate is measured as a change in concentration or activity as a function of
time. Unless the conditions of constant C_{sol} are specified (*e.g.*, a solution
flowing rapidly past a solid on a filter or the amount of solid dissolving
being so little as to leave C_{sol} unaffected or the time interval studied
being quite small) then the actual net flux of silica molecules per unit
surface area of solid varies as a function of the difference between C_{sol}
and C_{sat}. Thus, the net dissolution rate may vary between zero and
whatever maximum flux can be attained when $C_{sol} = 0$. In addition,
variable amounts of solid, and therefore variable values of S, can be put
into solution, further increasing or decreasing the dissolution rate while
the net flux per unit area at a given distance from equilibrium remains
the same. To compare the dissolution rate constants of different samples
also requires that the same order equation apply to the various forms

being studied unless reaction conditions are so stated as to compare net flux per unit area at both a given C_{sol} and similar congruent or incongruent solution patterns. The value of k_2 obtained by dissolving a small portion of an assemblage may not reflect the true composition of the material studied if there is a mixture of silica forms present; thus a small amount of silica gel mixed in with quartz gives a dC/dt curve which approaches quartz saturation much more rapidly than normal. The value of k_2 is usually determined by using the integral of Equation 4:

$$\ln \left(\frac{C_{sat} - C_{sol}}{C_{sat} - C_0} \right) = -k_2 St \qquad (5)$$

Since the actual value of C_{sat} is not used to determine k_2, but rather the rate at which the above fraction changes with time, C_{sat} may be numerically low but if approached in an artificially fast manner could yield a very large value of k_2. We believe this is the case for a number of the older samples which exhibit quite low solubilities but have moderate to high values of k_2 (*i.e.*, Nos. 27, 29, 34, 35, 36).

Consider for example a mixture of quartz and biogenic silica having the following properties at 25°C in pH 8.3 seawater as shown in Table II.

Even if we use a more conservative specific surface area value for radiolarians such as 200 cm^2 mg^{-1}, the initial concentration changes are still 500 times faster than those of the quartz, *i.e.*, only 0.2% by

Table III.   Measured and Calculated Properties of Acid-Cleaned

Sample Description, Depth in Core (cm)	No.	Age (mybp)	Specific Surface Area (m² gm⁻¹)	Upper Limit Solubility at 26°C (μM)	k₂ at 26°C (10⁻⁸ cm sec⁻¹)	S (cm² cm⁻³)
M–70, FFC2791, 10–30	32	Rec.	264	1647	10.2 ± 1.8	2640
M–70, FFC3693, 10–30	33	Rec.	248	1685	11.1 ± 1.2	2480
M–70–16, 500–523	4	2	71	1518	12 ± 1.3	400
S–68–33, 610–625	5	2	150	1615	2.8 ± .2	1500
M–70–7, 405–413	6	2	200	1583	3.2 ± .1	2000
M–70–39, 1474–1479	7	4.7	56	1630	14 ± 1.4	400
M–70–17, 300–325	8	4.6	150	1750	7.3 ± 1.6	1500
M–70–17, 470–490	11	6.4	190	1740	4.9 ± .7	1900
M–70–13, 310, 410, 357–364	12	6.4	88	1530	16.7 ± 1.7	400
M–70–17, 865–885	13	8	105	1720	7.3 ± .2	1050
M–70–13, 410–420	14	8	69	1360	14.2 ± 1.4	400
M–70–13, 505–515	16	10	75	1670	12 ± 1	400
M–70–17, 1015–1040	18	13	50	1600	7.4 ± .7	400
M–70–76, 1240–1250	20	15	53	1518	9.1 ± .9	400
M–70–10, 246–254	21	15	19.5	1200	18.3 ± 1	195
S–68–24, 220–230	36	16	8.5	300	16 ± 14	85
M–70–10, 991–1000	22	17.5	23	1280	25 ± 2.5	230
M–70–38, 650–660	23	17.5	33.4	1470	7.7 ± 1	400
S–68–24–425	38	17.5	39	775	6 ± 2.5	390
S–68–24, 690, 730, 740	25	18–20	16	520	6.4 ± .5	160
M–70–38, 1474–1479	24	20.5	22	1500	8.6 ± 1	400
S–68–24, 840–850	37	21	18.5	425	3 ± 2	185
S–68–24, 1000–1008	26	22–23	56.7	1335	14 ± 2.5	567
S–68–24, 1320	27	25	21	790	22.6 ± 2	210
KK–72–39, 715, 765	31	28–29	39	1483	11.8 ± 1.7	390
KK–72–39, 1215, 1265	30	34–38	56	1420	11.6 ± 1.4	560
S–68–24, 1637–1640, 1650	29	34–38	20	660	36 ± 4	200
S–68–24, 2091–2100, 2120	28	40	7.1	245	—	71
S–68–24, 2145	34	40	2.2	100	146 ± 160	22
M–70–39, 2080	35	40	20.8	625	12 ± 6	208

Table II.   Properties of a Mixture of Quartz and Biogenic Silica

	Quartz	*Biogenic Silica*
k_2 (cm sec^{-1})	2×10^{-8}	10×10^{-8}
Specific surface area (cm^2 mg^{-1})	20	2000
S (cm^2 cm^{-3})	20	2000
C_{sat} (mole cm^{-3})	1×10^{-7}	20×10^{-7}
Flux from surface of solid (moles cm^{-2} sec^{-1})	2×10^{-15}	100×10^{-15}
Initial dC/dt (mole cm^{-3} sec^{-1})	4×10^{-14}	$40{,}000 \times 10^{-14}$

weight contamination of the lower specific surface area radiolarians will double the initial value of dC/dt. These calculations suggest that the problems associated with having a mixture of silica forms in older samples may be a major drawback in attempting to use k_2 as a direct indicator of silica form change.

For the above reasons it is sometimes difficult to judge what value of C_{sat} should be used for a given assemblage. We have arbitrarily chosen to use the highest value of dissolved silica obtained during a particular run for calculations involving that sample. In the case of older samples this tends to overestimate the value of C_{sat} because of the mixture of silica forms present. Possibly a more viable technique for older specimens would be that used by Stöber (*15*) and Baumann (*31*)

Biogenic Silica from the Central Equatorial Pacific

$\Delta G°$ at $26°C$ (kcal/mole)	$\Delta G‡$ at $26°C$ (kcal/mole)	Upper Limit Solubility at $3°C$ (μM)	k_2 at $3°C$ (10^{-8} cm sec^{-1})	S, (cm^2 cm^{-3})	ΔG at $3°C$, (kcal/mole)	$\Delta G‡$ at $3°C$, (kcal/mole)	$\Delta H‡ + RT$ at $3°-26°C$	ΔH at $3°-26°C$
3.81	14.9	959	0.60 ± 0.06	2640	3.81	15.3	19.1	3.86
3.79	14.9	970	0.60 ± 0.06	2480	3.81	15.3	20.8	3.94
3.86	14.8	1004	1.3 ± 0.9	710	3.79	14.9	15.8	2.95
3.82	15.7	861	0.24 ± 0.02	1500	3.87	15.8	17.5	4.48
3.83	17.9	887	0.26 ± 0.03	2000	3.85	15.8	17.9	4.13
3.81	14.8	849	0.4 ± 0.04	560	3.88	15.5	25.3	4.65
3.77	15.1	995	0.6 ± 0.06	1500	3.79	15.3	17.8	4.03
3.77	15.3	927	0.36 ± 0.04	1900	3.83	15.6	18.6	4.49
3.84	14.6	981	0.97 ± 0.1	880	3.80	15.1	18.8	3.71
3.78	15.1	1006	0.7 ± 0.07	1050	3.79	15.2	18.7	3.82
3.92	14.7	991	1.2 ± 0.1	690	3.79	14.9	17.6	2.26
3.80	14.8	957	0.94 ± 0.09	750	3.81	15.1	18.1	3.97
3.81	15.1	825	0.47 ± 0.05	500	3.89	15.5	19.6	4.72
3.86	14.9	820	0.70 ± 0.07	530	3.90	15.2	18.3	4.39
4.00	14.5	740	1.5 ± 0.15	195	3.95	14.8	17.8	3.47
4.80	14.6	69	—	85	—	—	5.2	—
3.94	14.3	844	1.4 ± 0.14	23	3.92	14.9	20.5	3.44
3.90	15.1	836	0.6 ± 0.06	334	3.92	15.3	17.1	4.52
4.24	15.2	592	—	390	—	—	6.5	—
4.49	15.2	389	—	160	—	—	—	—
3.85	15.0	788	0.95 ± 0.09	220	3.98	15.1	15.7	5.43
4.60	15.6	218	—	185	—	—	—	—
3.92	14.7	808	0.97 ± 0.09	567	3.95	15.1	19.0	4.11
4.24	14.4	693	1.8 ± 0.2	210	4.05	14.7	18.0	1.67
3.87	14.8	840	0.95 ± 0.09	390	3.92	15.1	17.9	4.49
3.88	14.8	848	1.1 ± 0.1	560	3.88	15.0	16.8	3.68
4.34	14.1	678	1.8 ± 0.2	200	4.12	14.7	21.3	1.30
4.94	—	239	3.3 ± 0.3	71	—	14.4	—	—
5.45	13.3	573	—	22	—	—	14.2	—
4.35	14.8	443	—	208	—	—	7.8	—

of successive extractions followed by redetermination of specific surface area after a constant solution rate was attained. We are now preparing to do this on several of our older samples. Younger samples, with their higher dissolved silica concentrations, have the difficulty of producing silica concentrations which are capable of precipitating a poorly ordered magnesium–hydroxyl–silicate, as shown by Wollast *et al.* (*32*), thereby additionally complicating reaction kinetics.

The value of k_2 increases with increasing temperature, pH, and ionic strength for a given form of silica (*1, 2, 9, 22*). A major difficulty arises when unbuffered solutions of differing ionic strength are compared. As Stumm and Morgan (*33*) pointed out, the pH at the surface of a solid in low-ionic-strength aqueous solutions may be at least 0.9 units less than the bulk solution value while the surface of the same solid in seawater may be only 0.2 pH units less. Since Hurd (*2*) has shown that k_2 for biogenic silica varies by a factor of 2–3 in the range pH 7.3–8.3, such factors must be considered before the effect of a given variable such as ionic strength can be understood. All solutions must be well buffered.

With the above limitations in mind, the data in Table IV are presented. These values were calculated from the few articles in the litera-

Table IV.   Selected Kinetic and Thermodynamic Properties of Several Forms of Silica

	Quartz—25°C, pH 8.3–8.5		
	Distilled Water	*9% NaCl*	*Seawater (est.)*
k_2	5.8×10^{-9} cm sec^{-1},	1–2×10^{-8} cm sec^{-1},	2–3×10^{-8} cm sec^{-1},
$\Delta G_{k2}^{\ddagger}$	16.6 kcal/mole (Baumann)	15.8–16.2 kcal/mole	15.6–15.8 kcal/mole
k_2	3.7×10^{-9} cm sec^{-1},	(Stöber)	
$\Delta G_{k2}^{\ddagger}$	16.8 kcal/mole (Van Lier, estimated and unbuffered) recalculated from higher temperatures		
	Cristobalite—25°C, pH 8.5		
k_2	—	$.6$–1.3×10^{-8} cm sec^{-1},	2–3×10^{-8} cm sec^{-1},
$\Delta G_{k2}^{\ddagger}$		16.2–16.6 kcal/mole (Stöber)	16.2 kcal/mole
	Vitreous Silica		
k_2	2×10^{-9} cm sec^{-1},	2×10^{-8} cm sec^{-1},	~3–4×10^{-8} cm sec^{-1},
$\Delta G_{k2}^{\ddagger}$	17.2 kcal/mole (Baumann by comparison with quartz in other solutions)	15.8 kcal/mole (Stöber)	15.4–15.8 kcal/mole
	Silica Gel		
k_2	4.5×10^{-9} cm sec^{-1},	—	5–15×10^{-8} cm sec^{-1},
$\Delta G_{k2}^{\ddagger}$	16.7 kcal/mole (recalculated from higher temperatures, pH unknown) (Greenberg)		14.6–15.4 kcal/mole
	Biogenic Silica		
k_2	—	—	2.8–25×10^{-8} cm sec^{-1} mean value 9.6×10^{-8} cm sec^{-1}. standard deviation 5.6×10^{-8} cm sec^{-1},
$\Delta G_{k2}^{\ddagger}$			14.9 ± 0.5 kcal/mole

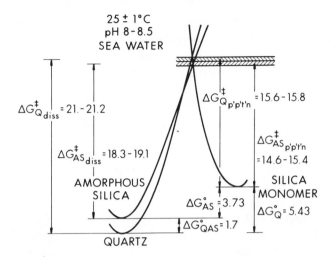

Figure 11. Reaction coordinate describing the relative amounts of energy required by quartz and amorphous silica to dissolve and precipitate in 25° ± 1°C, pH 8–8.5 seawater. Energy changes are in kcal mole⁻¹ and are given as absolute values.

ture that contain enough information to estimate a rate constant. Methods for the calculation of the rate constants from the authors' data are given in Appendix III. Experimental methods and an illustration showing several typical dC/dt plots are given in Appendix I. (Raw data available upon request)

Figure 11 uses the data in Tables I and IV to explain the energy relationships among two of the different forms of silica, the hydrated silica monomer, and their respective activated complexes. In this illustration, the term amorphous silica is synonymous with the vitreous silica–silica gel grouping mentioned earlier. The relative free energy positions of the two solid forms of silica and their activated complexes were plotted relative to the position of the hydrated silica monomer in seawater, pH 8–8.5 at 25 ± 1°C. The equations involving the calculation of various $\Delta G°$ and $\Delta G‡$ values are given in Appendix II. In brief, however, the driving force behind form changes *per se* (amorphous silica to cristobalite or quartz) is the net difference in free energy of the forms as calculated from their solubilities. The rate at which these processes occur is a function of the free energy of activation which is directly obtainable from the heterogeneous rate constant by the equation (*34*):

$$k_2 = (RT/2\pi W)^{1/2} \exp\left(-\Delta G‡/RT\right) \tag{6}$$

where the value of R in the first parenthesis is 8.31×10^7 erg mole⁻¹ deg K^{-1} and that in the second parenthesis is 1.987 cal mole⁻¹ deg K^{-1}. W is the molecular weight of the activated complex, and the other variables are as previously described. When $R(\ln k_2)$ is plotted as a function of the reciprocal of the absolute temperature, the slope of the line is the value of the enthalpy of activation, $\Delta H^{\ddagger}$ plus RT (24). Unless the entropy of activation is known by some other means for the process studied, the driving force behind the rate at which the reaction occurs, $\Delta G^{\ddagger}$, cannot be estimated simply by knowing the enthalpy of activation. This emphasizes the importance of the determination of the values of k_2 for different forms of silica under standard and reproducible sets of conditions before comparisons can be made of changes between phases and forms. It further shows the importance of increasing pH and ionic strength in reducing the value of $\Delta G^{\ddagger}$ by increasing the numerical value of k_2. Experiments which are performed at much higher temperatures to reduce reaction completion times may thus be subject to serious errors; an order of magnitude increase in k_2 produces an energy change of $-RT\ln(9.9 \times 10^{-2})$ in the value of $\Delta G^{\ddagger}$.

Figure 11 shows that greater free energy changes are required for both the solution and precipitation of quartz than for amorphous silica, based simply on the differences between their solubilities and solution rate constants. This may account in part for the frequent supersaturation of quartz solutions; silica monomer can absorb to the quartz surface without actually becoming a part of the quartz structure. This effect is enhanced in solutions of decreasing ionic strength and pH where $\Delta G^{\ddagger}$ increases as k_2 decreases.

Appendix I

Several cores from the Central Equatorial Pacific which were previously dated (36, 37) were sampled at various depths to provide a continuous series of samples ranging in age from Recent to Late Eocene (40 mybp).

Each assemblage was cleaned first by sieving with a 62μ mesh screen, then heating $(60°-70°C)$ first in dilute H_2O_2, then in dilute HCl for several hours, followed by resieving, washing with distilled water, and drying overnight at $100°-105°C$. The specific surface area of each assemblage was then analyzed by nitrogen adsorption (2). The samples were dissolved in tris-hydroxymethylaminomethane-buffed, pH 8.3, surface seawater at $26° \pm 1°C$ and $3° \pm 1°C$ as described earlier (1, 2).

Unless otherwise noted each dissolution experiment contained 50 mg of solid in 50 cm^3 of solution. The value of S for each run is given in Table III. Dissolved silica was determined as in Hurd (2). How long one must wait for equilibrium to be established depends on what form of silica is present and what the value for S for that experiment is. Given

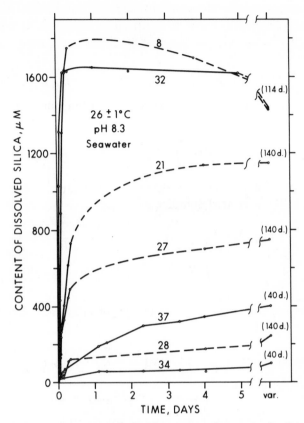

Figure 1A.   Typical dC/dt *curves for several of the samples studied*

S, however, and using a value for k_2 representative of amorphous silica, one can calculate the earliest that equilibrium will be established by using Equation 5. In this equation, simply substitute the fraction completion desired (*i.e.*, 0.05 for 95%, 0.001 for 99.9%, etc.), insert the appropriate values for S and k_2, and solve for t. For values of S between 20 and 2000 and a k_2 of 1×10^{-7} cm sec^{-1} for seawater at 25°C, pH 8.3, completion to 95% of equilibrium will be attained within 4 hr–17 days.

Figure 1A shows several typical dC/dt curves from which values of k_2 and other properties were calculated. (Raw data available upon request.)

Appendix II

The following argument follows the logic presented by O'Connor and Greenberg (*27*) and Van Lier *et al.* (*22*) wherein the ratio of the foreward and reverse rate constants for the interaction of quartz or amorphous silica with water to form silica monomer is equal to the equilibrium constant of that given form. Thus:

$$K_{eq(quartz)} = \frac{k_{1(quartz)}}{k_{2(quartz)}}; \quad K_{eq(amorph.\ sil.)} = \frac{k_{1(amorph.\ sil.)}}{k_{2(amorph.\ sil.)}}$$

and, since the change in free energy going from quartz or amorphous silica to silica monomer is:

$$K_{eq(quartz)} = \exp\left(\frac{-\Delta G^{\circ}{}_{(quartz)}}{RT}\right); \quad K_{eq(amorph.\ sil.)} = \exp\left(\frac{-\Delta G^{\circ}{}_{(amorph.\ sil.)}}{RT}\right)$$

and since the change in free energy of activation going from silica monomer to the activated complex is (34):

$$k_{2(quartz)} = \left(\frac{RT}{2\pi\omega}\right)^{1/2} \exp\left(\frac{-\Delta G^{\ddagger}{}_{(quartz\ ppt.)}}{RT}\right);$$

$$k_{2(amorph.\ sil.)} = \left(\frac{RT}{2\pi\omega}\right)^{1/2} \exp\left(\frac{-\Delta G^{\ddagger}{}_{(amorph.\ sil.\ ppt.)}}{RT}\right)$$

Therefore, the change in free energy of activation of dissolution as shown in Figure 11 is:

$$k_1 = k_2 \cdot K_{eq} = \left(\frac{RT}{2\pi\omega}\right)^{1/2} \exp\left(\frac{-\Delta G^{\ddagger}{}_{(ppt)}}{RT}\right) \exp\left(\frac{-\Delta G^{\circ}}{RT}\right) \quad \text{or}$$

$$\Delta G^{\ddagger}{}_{(dissolution)} = \Delta G^{\ddagger}{}_{(ppt.)} + \Delta G^{\circ}$$

and by a similar argument, the ratio of the reaction rates must equal:

$$\frac{k_{1(amorph.\ sil.)}}{k_{1(quartz)}} = \exp\left(\frac{-\Delta G^{\ddagger}{}_{(amorph.\ sil.\ diss.)} + \Delta G^{\ddagger}{}_{(quartz\ diss.)}}{RT}\right)$$

$$= 25 \text{ to } 135$$

Appendix III

A general method for determining a value of k_2 from a given experiment such as the repetitive extraction of a solid by a solvent for several weeks (15, 31) is to use the following equation:

$$\ln\left(\frac{C_{sat} - C_{sol}}{C_{sat} - C_0}\right) = -k_2 St$$

and assume the following:

1. C_{sol} is the molar concentration of the dissolved solid in solution at time t; i.e., $5 \times 10^{-6} M$ after 1 day (or 86,400 sec.).

2. C_{sat} is the saturation value of the form studied at a given temperature and pH.

3. C_0 is the concentration of dissolved silica in solution when $t = 0$.

4. S is the surface area of the solid per unit volume of solution in cm^2 cm^{-3}.

Example: in Stöber (*15*), using Figure 15, p. 177, for vitreous silica:

$S = 200$ cm^{-1}, $C_{sol} = 39$ ppm, $C_{sat} = 120$ ppm, $C_0 = 0$, $t = 86,400$ sec

$$k_2 = \frac{\ln\left(\dfrac{120 - 39}{120 - 0}\right)}{(200 \text{cm}^{-1})(8.64 \times 10^4 \text{ sec})}$$

$$= -2.3 \times 10^{-8} \text{cm sec}^{-1} \text{ or}$$
$$1.36 \times 10^{-6} \text{ furlong fortnight}^{-1}$$

Acknowledgments

The authors gratefully acknowledge the technical assistance of the following people: Janet Olmon for her tireless and accurate silicate and surface area analyses; Glen Sicks for a portion of the surface area determinations; Dennis T. O. Kam for computer programming services; Arthur Hubbard, Chemistry Department, University of Hawaii, for suggesting the use of the free energy of activation equations; and the Hawaii Institute of Geophyhics Core Laboratory for searching the bowels of the core lockers for our samples.

Literature Cited

1. Hurd, D. C., *Earth Planet. Sci. Lett.* (1972) **15**, 411.
2. Hurd, D. C., *Geochim. Cosmochim. Acta* (1973) **37**, 2257.
3. Jones, M. M., Pytkowicz, R. M., *Bull. Soc. Sci. Liège* (1973) **42**, 118.
4. Kamatani, A., *J. Ocean. Soc. Japan* (1969) **25**, 1.
5. Lewin, J. C., *Geochim. Cosmochim. Acta* (1961) **21**, 182.
6. Alexander, G. B., Heston, W. M., Iler, R. K., *J. Phys. Chem.* (1954) **58**, 453.
7. Greenberg, S. A., *J. Phys. Chem.* (1957) **61**, 196.
8. Iler, R. K., "Colloid Chemistry of Silica and Silicates," Cornell University, Ithaca, N.Y., 1955.
9. Kitahara, S., Ooshima, F., *Nippon Kagaku Zasshi* (1966) **87**, 316.
10. Krauskopf, K. B., *Geochim. Cosmochim. Acta* (1956) **10**, 1.
11. Krauskopf, K. B., *Soc. Econ. Paleontol. Mineral. Spec. Publ.* (1959) **7**, 4.
12. Morey, G. W., Fournier, R. O., Rowe, J. J., *J. Geophys. Res.* (1964) **69**, 1995.
13. Okamoto, G., Okura, T., Goto, K., *Geochim. Cosmochim. Acta* (1957) **10**, 123.
14. Siever, R., *J. Geol.* (1962) **70**, 127.
15. Stöber, W., ADV. CHEM. SER. (1967) **67**.
16. White, D. E., Brannock, W. W., Murata, *Geochim. Cosmochim. Acta* (1956) **10**, 27.
17. Sosman, R. B., "The Phases of Silica," 388 pp., Rutgers University, New Brunswick, 1965.

18. Fournier, R. O., Rowe, J. J., *Am. Mineral.* (1962) **47**, 897.
19. Morey, G. W., Fournier, R. O., Rowe, J. J., *Geochim. Cosmochim. Acta* (1962) **26**, 1029.
20. Fournier, R. O., *Proc. Int. Symp. Hydrogeochem. Biogeochem., Tokyo, 1970* (1973) 122.
21. Kennedy, G. C., *Econ. Geol.* (1950) **45**, 629.
22. Van Lier, J. A., deBruyn, P. L., Overbeek, J. Th. G., *J. Phys. Chem.* (1960) **64**, 1675.
23. Mackenzie, F. T., Gees, R., *Science* (1971) **173**, 533.
24. Daniels, F., Alberty, R. A., "Physical Chemistry," 744 pp., John Wiley & Sons, New York, 1961.
25. Alexander, G. B., *J. Phys. Chem.* (1957) **61**, 1563.
26. Harder, H., *Mineral. Soc. Japan. Spec. Paper* (1971) **1**, 106.
27. O'Connor, T. L., Greenberg, S. A., *J. Phys. Chem.* (1958) **62**, 1195.
28. Grill, E., Richards, F., *J. Mar. Res.* (1964) **22**, 51.
29. Kamatani, A., *Mar. Biol.* (1971) **8**, 89.
30. Kato, K., Kitano, Y., *J. Ocean. Soc. Japan* (1968) **24**, 147.
31. Baumann, H., *Beitr. Silikose–Forsch.* (1965) **85**, 1.
32. Wollast, R., Mackenzie, F. T., Bricker, O. P., *Am. Mineral.* (1968) **53**, 1945.
33. Stumm, W., Morgan, J., "Aquatic Chemistry," 583 pp., Wiley-Interscience, New York, 1970.
34. Lai, C. N., Hubbard, A. T., *Inorg. Chem.* (1974) **13**, 1199.
35. Wollast, R., "The Sea, Marine Chemistry," E. Goldberg, Ed., Vol. 5, 895 pp., Wiley-Interscience, New York, 1974.
36. Theyer, F., Hammond, S., *Earth Planet. Sci. Lett.* (1974a) **22**, 307.
37. Theyer, F., Hammond, S., *Geology* (1974b) **2**, 487.
38. Lerman, A., Mackenzie, F. T., Bricker, O. P., *Earth Planet. Sci. Lett.* (1957) **25**, 82.

RECEIVED January 3, 1975. This work was supported by Office of Naval Research Contract N00014-70-A-0016-0001. This paper is Contribution No. 639 at the Hawaii Institute of Geophysics.

INDEX

INDEX

233

The text of this book is set in 10 point Caledonia with two points of leading. The chapter numerals are set in 30 point Garamond; the chapter titles are set in 18 point Garamond Bold.

The book is printed offset on Text White Opaque, 50-pound. The cover is Joanna Book Binding blue linen.

Jacket design by Diane Reich.
Editing and production by Virginia Orr.

The book was composed by the Mills-Frizell-Evans Co. and by Service Composition Co., Baltimore, Md.; printed and bound by The Maple Press Co., York, Pa.